普通高等院校电子信息系列教材

|OpenCL|
异构计算
入门FPGA和TensorFlow神经网络

胡正伟 谢志远 王岩 ◎编著

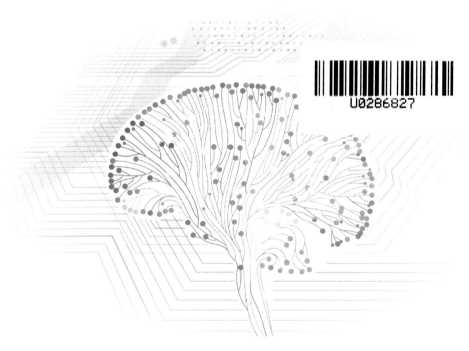

清华大学出版社
北京

内容简介

本书以通过FPGA实现简易神经网络的推理流程为主线,主要包含以下内容:在TensorFlow学习框架下实现神经网络训练,保存训练好的权值和偏置;将TensorFlow框架下训练的神经网络使用OpenCL语言实现,并编译生成可执行文件和FPGA编程文件;将输入数据、权值、偏置等数据通过以太网口传输到FPGA开发板;在FPGA开发板上运行神经网络。

本书的重点在于神经网络算法的OpenCL描述方法及FPGA实现流程。简易神经网络算法不仅可以让读者明白神经网络的工作原理及基本框架,还可以使用较少的OpenCL代码描述,易于分析神经网络算法与代码的对应关系,实现OpenCL语言的学习。

本书以Ubuntu操作系统为运行环境,以性价比高的FPGA开发板DE10_nano为实现平台,该开发板尺寸较小,易于携带,方便管理,价格较低,适合批量购买以开展相关教学实验。

本书面向电子信息、计算机、自动化等相关专业的本科生及研究生或FPGA开发人员。

本书封面贴有清华大学出版社防伪标签,无标签者不得销售。
版权所有,侵权必究。举报:010-62782989,beiqinquan@tup.tsinghua.edu.cn。

图书在版编目(CIP)数据

OpenCL异构计算:入门FPGA和TensorFlow神经网络/胡正伟,谢志远,王岩编著. ―北京:清华大学出版社,2021.12(2024.11重印)
普通高等院校电子信息系列教材
ISBN 978-7-302-59398-0

Ⅰ.①O… Ⅱ.①胡… ②谢… ③王… Ⅲ.①可编程序逻辑器件-系统设计-高等学校-教材 ②人工智能-算法-高等学校-教材 Ⅳ.①TP332.1 ②TP18

中国版本图书馆CIP数据核字(2021)第212820号

责任编辑:郭 赛
封面设计:杨玉兰
责任校对:徐俊伟
责任印制:宋 林

出版发行:清华大学出版社
    网　　址:https://www.tup.com.cn,https://www.wqxuetang.com
    地　　址:北京清华大学学研大厦A座　　邮　编:100084
    社 总 机:010-83470000　　邮　购:010-62786544
    投稿与读者服务:010-62776969,c-service@tup.tsinghua.edu.cn
    质量反馈:010-62772015,zhiliang@tup.tsinghua.edu.cn
    课件下载:https://www.tup.com.cn,010-83470236
印 装 者:三河市人民印务有限公司
经　　销:全国新华书店
开　　本:185mm×260mm　　印 张:14.75　　字 数:358千字
版　　次:2021年12月第1版　　印 次:2024年11月第3次印刷
定　　价:59.00元

产品编号:092775-01

前言
Foreword

为了便于开展面向 FPGA 平台的 OpenCL 教学,本书以人工智能领域中的神经网络为实现目标,通过学习 TensorFlow 框架下的神经网络训练、神经网络算法的 OpenCL 描述、神经网络的 FPGA 实现等内容,帮助读者掌握 FPGA 实现神经网络算法推理的整个流程,为今后从事人工智能、算法加速、FPGA 开发等相关领域的工作奠定基础。

鉴于作者水平,本书未能详尽介绍 OpenCL 的优化策略以及复杂神经网络算法的实现,作者力争在后续版本中完成相关内容的介绍。

欢迎本领域的相关专家学者、读者批评指正,希望本书能够起到抛砖引玉的效果。

本书的出版受到了华北电力大学"双一流"研究生教材项目及国家自然科学基金项目(52177083)的支持。

本书学习指导

(1) 学习重点是基于 OpenCL 的 FPGA 开发流程。

(2) 书中未对 API 函数(TensorFlow API、OpenCL API)进行语法知识介绍,若想深入了解相关 API 的参数意义及详尽使用,请利用网络查询或查看数据手册。

(3) 请自行学习 Ubuntu 操作系统的使用方法并熟练掌握基本命令。

(4) 读者可采用不同的 Quartus Prime 版本。

(5) 读者可采用其他 FPGA 开发板,但需要结合配套的 BSP。

(6) 读者需根据实际的路径信息修改本书例子中的路径信息。

(7) 本书的参考学时为 32 学时,建议采用理论实践一体化的教学模式,各章的参考学时详见如下学时分配表。

学时分配表

项目及章节	课 程 内 容	学时
第 1 章	绪论	2
第 2 章	TensorFlow 基础知识及运行环境搭建	2
实验 1	TensorFlow 基础命令	1
第 3 章	TensorFlow 实现神经网络模型训练与测试	2

续表

项目及章节	课程内容	学时
实验 2	TensorFlow 实现简易神经网络模型的训练与测试	1
实验 3	TensorFlow 实现卷积神经网络模型的训练与测试	1
实验 4	TensorFlow 实现 MNIST 数据集转换	1
实验 5	读取 tfrecords 格式数据并实现 MNIST 手写字体识别	1
第 4 章	OpenCL 基础	2
第 5 章	面向 Intel FPGA 的 OpenCL 运行平台搭建	2
实验 6	DE10_nano 开发板运行 OpenCL 程序	1
实验 7	DE10_nano 与 PC 交换数据	1
实验 8	OpenCL 程序编译	1
实验 9	编写第一个 OpenCL 程序	1
第 6 章	单层神经网络算法模型的 FPGA 实现流程	2
实验 10	单层神经网络算法模型的 FPGA 实现流程	2
第 7 章	单层神经网络算法的 kernel 程序实现方式分析比较	2
实验 11	单层神经网络算法的 kernel 程序的不同实现方式	1
第 8 章	具有一个隐形层的神经网络算法模型的 OpenCL 实现	2
实验 12	具有一个隐形层的神经网络算法模型的 OpenCL 实现	1
第 9 章	简易卷积神经网络的 OpenCL 实现	2
实验 13	简易卷积神经网络算法模型的 OpenCL 实现	1

扫描下方二维码，即可加入本教材的 QQ 交流群。

编 者

2021 年 11 月

目录

第1章 绪论 ·········· 1
1.1 异构计算系统 ·········· 1
1.2 OpenCL ·········· 2
1.3 FPGA ·········· 3
1.4 FPGA+CPU 异构计算系统 ·········· 5
1.5 HDL 和 OpenCL ·········· 6
    1.5.1 OpenCL 的优点 ·········· 7
    1.5.2 OpenCL 的缺点 ·········· 7
1.6 人工神经网络 ·········· 8
    1.6.1 人工神经网络的基本概念 ·········· 8
    1.6.2 人工神经网络的基本特征 ·········· 9
    1.6.3 人工神经网络的应用 ·········· 10
习题 1 ·········· 12

第2章 TensorFlow 基础知识及运行环境搭建 ·········· 14
2.1 TensorFlow 简介 ·········· 14
2.2 TensorFlow 两步编程模式 ·········· 14
2.3 TensorFlow 两步编程模式实例 ·········· 15
    2.3.1 定义计算图的基本操作 ·········· 15
    2.3.2 运行计算图的基本操作 ·········· 18
2.4 TensorFlow 环境搭建 ·········· 23
    2.4.1 软件安装 ·········· 23
    2.4.2 TensorFlow 软件运行 ·········· 25
    2.4.3 计算图例程运行实例 ·········· 25
习题 2 ·········· 30

第 3 章 TensorFlow 实现神经网络模型训练与测试 ···································· 31
3.1 神经网络训练与测试的基本概念 ·· 31
3.1.1 神经网络的训练 ·· 31
3.1.2 神经网络的测试 ·· 32
3.2 基于 TensforFlow 训练神经网络实现 MNIST 数据集识别 ··························· 32
3.2.1 MNSIT 数据集 ·· 32
3.2.2 Softmax Regression 模型 ··· 33
3.2.3 MNIST 数据识别的 Softmax Regression 神经网络模型 ······························ 35
3.2.4 MNIST 数据识别的卷积神经网络模型 ·· 40
3.3 MNIST 数据集转换 ·· 49
3.3.1 将数据集转换为以 txt 文件保存的数据 ··· 49
3.3.2 将数据集转换为以 bmp 文件保存的图片 ··· 50
3.3.3 将 bmp 转换为 tfrecords 格式 ·· 54
3.4 读取 tfrecords 格式数据实现 MNIST 手写字体识别 ···································· 56
3.4.1 Softmax Regression 模型 ··· 56
3.4.2 卷积神经网络模型 ··· 58
习题 3 ··· 61

第 4 章 OpenCL 基础 ··· 63
4.1 OpenCL 标准框架 ·· 63
4.2 OpenCL 基本概念基础 ··· 64
4.3 OpenCL 程序的组成部分 ·· 65
4.4 OpenCL 框架的 4 种模型 ·· 66
4.5 编写第一个 OpenCL 程序 ··· 71
4.5.1 kernel 程序 ·· 71
4.5.2 host 程序 ··· 72
4.6 OpenCL 基本知识点 ··· 78
4.6.1 kernel 函数格式 ··· 78
4.6.2 kernel 编程模式 ··· 79
4.6.3 kernel 地址限定符 ·· 79
4.6.4 kernel 语句描述 ··· 80
4.6.5 kernel 数据类型 ··· 80
4.6.6 kernel 编程限制 ··· 80
习题 4 ··· 80

第 5 章 面向 Intel FPGA 的 OpenCL 运行平台搭建 ·· 82
5.1 搭建 OpenCL 平台的软硬件要求 ·· 82
5.2 面向 OpenCL 应用的 DE10_nano 开发板简介 ··· 83

5.3 平台所需软件下载 ……………………………………………………………… 84
　　5.3.1 Quartus Prime Standard 下载 …………………………………………… 84
　　5.3.2 Intel FPGA SDK for OpenCL 下载 ……………………………………… 85
　　5.3.3 Intel SoC FPGA EDS 下载 ……………………………………………… 85
5.4 平台所需软件安装 ……………………………………………………………… 86
　　5.4.1 安装 Quartus Prime Standard Edition＋Intel FPGA SDK for
　　　　　OpenCL …………………………………………………………………… 86
　　5.4.2 安装 SoCEDS …………………………………………………………… 91
　　5.4.3 安装 DE10_nano BSP …………………………………………………… 95
5.5 环境变量设置 …………………………………………………………………… 96
　　5.5.1 环境变量设置步骤 ……………………………………………………… 96
　　5.5.2 环境变量测试 …………………………………………………………… 97
5.6 编译 OpenCL kernel …………………………………………………………… 98
5.7 编译 host 程序 ………………………………………………………………… 98
5.8 烧写 img 文件到 SD 卡（在 Windows 系统下完成）………………………… 99
5.9 minicom 驱动安装与测试 ……………………………………………………… 101
　　5.9.1 minicom 驱动安装 ……………………………………………………… 101
　　5.9.2 minicom 使用测试 ……………………………………………………… 102
5.10 hello world kernel 运行测试 ………………………………………………… 103
5.11 DE10_nano 与 PC 交换数据 ………………………………………………… 104
习题 5 ………………………………………………………………………………… 108

第 6 章 单层神经网络算法模型的 FPGA 实现流程 ……………………………… 109
6.1 基于 OpenCL 的神经网络算法设计与 FPGA 实现的基本流程 …………… 109
6.2 无隐形层的简易神经网络算法原理 …………………………………………… 110
6.3 神经网络的 TensorFlow 实现及训练 ………………………………………… 111
6.4 TensorFlow 框架下输入数据的转换 ………………………………………… 114
6.5 神经网络算法的 OpenCL 实现 ………………………………………………… 115
　　6.5.1 kernel 代码编写及编译 ………………………………………………… 115
　　6.5.2 host 代码编写及编译 …………………………………………………… 116
6.6 数据移植复制到 FPGA 开发板 ………………………………………………… 120
6.7 FPGA 运行神经网络 …………………………………………………………… 123
6.8 kernel report.html 文件查看 ………………………………………………… 124
　　6.8.1 高层设计报告布局 ……………………………………………………… 124
　　6.8.2 系统概要 ………………………………………………………………… 125
　　6.8.3 迭代分析 ………………………………………………………………… 127
　　6.8.4 资源分析 ………………………………………………………………… 128
　　6.8.5 系统视图 ………………………………………………………………… 131

6.9　log 文件查看 FPGA 资源使用估计信息 ·· 133
习题 6 ·· 133

第 7 章　单层神经网络算法的 kernel 程序实现方式分析比较 ······················ 135
7.1　批量读取输入数据的 OpenCL 程序 ·· 135
  7.1.1　kernel 程序 ··· 135
  7.1.2　host 程序 ··· 136
  7.1.3　执行结果 ··· 142
7.2　神经网络算法的不同 kernel 代码实现对比 ·· 142
  7.2.1　single work item 和 NDRange(private) ·· 142
  7.2.2　local 和 private(single work item) ·· 145
  7.2.3　local 和 private(NDRange) ·· 148
  7.2.4　single work item 和 NDRange(local) ··· 150
  7.2.5　float 和 char(single work item-local) ··· 154
  7.2.6　float 和 char(NDRange-private) ·· 156
7.3　神经网络算法的 ARM 与 FPGA 实现方式对比 ·· 159
  7.3.1　ARM 和 FPGA(float 数据类型) ··· 159
  7.3.2　ARM 和 FPGA(char 数据类型) ··· 162
7.4　host 代码与 kernel 的对应 ··· 165
习题 7 ·· 165

第 8 章　具有一个隐形层的神经网络算法模型的 OpenCL 实现 ······················ 166
8.1　一个隐形层的简易神经网络算法原理 ··· 166
8.2　具有一个隐形层的神经网络的 TensorFlow 实现及训练 ································· 168
8.3　具有一个隐形层的神经网络算法的 OpenCL 实现 ·· 171
  8.3.1　ARM 实现 ·· 171
  8.3.2　single work item 格式,一个 kernel ··· 171
  8.3.3　NDRange 格式,一个 kernel ··· 174
  8.3.4　single work item 格式,两个 kernel ··· 176
  8.3.5　NDRange 格式,两个 kernel ··· 178
  8.3.6　single work item 格式,两个 kernel,channel ···································· 181
  8.3.7　single work item 格式,两个 kernel,pipe ······································· 182
习题 8 ·· 186

第 9 章　简易卷积神经网络的 OpenCL 实现 ·· 187
9.1　简易卷积神经网络算法结构与原理 ·· 187
9.2　简易卷积神经网络的 TensorFlow 实现及训练 ·· 189
9.3　简易卷积神经网络算法的 OpenCL 实现 ··· 194

        9.3.1 NDRange 实现 ··· 194
        9.3.2 single work item 实现 ··· 206
   习题 9 ·· 218

第 10 章 上机实验 ·· 219
   实验 1   TensorFlow 基础命令 ··· 219
   实验 2   TensorFlow 实现简易神经网络模型的训练与测试 ······························ 219
   实验 3   TensorFlow 实现卷积神经网络模型的训练与测试 ······························ 220
   实验 4   TensorFlow 实现 MNIST 数据集转换 ··· 220
   实验 5   读取 tfrecords 格式数据并实现 MNIST 手写字体识别 ························· 220
   实验 6   DE10_nano 开发板运行 OpenCL 程序 ··· 220
   实验 7   DE10_nano 与 PC 数据交换 ··· 221
   实验 8   OpenCL 程序编译 ·· 221
   实验 9   编写一个 OpenCL 程序 ·· 221
   实验 10 单层神经网络算法模型的 FPGA 实现流程 ······································ 221
   实验 11 单层神经网络算法的 kernel 程序的不同实现方式 ······························ 221
   实验 12 具有一个隐形层的神经网络算法模型的 OpenCL 实现 ······················· 222
   实验 13 简易卷积神经网络算法模型的 OpenCL 实现 ··································· 223

参考文献 ·· 224

第 1 章 绪 论

本章首先引入异构计算系统的概念,然后介绍可降低异构计算系统开发难度的 OpenCL 以及高能效比的异构设备 FPGA,分析对比 OpenCL 与 HDL 开发 FPGA 异构系统的优缺点,最后介绍异构计算系统的典型应用领域——人工神经网络。

1.1 异构计算系统

系统智能化的提高离不开密集的计算量需求,当单个处理器系统不能胜任计算需求时,就会继而诞生多个并行计算平台以实现计算力的提升。异构计算系统使用几种不同类型的处理器优化执行计算任务。在异构计算系统中,不同类型的处理器,如中央处理器(CPU)、数字信号处理器(DSP)、图形处理单元(GPU)、现场可编程门阵列(FPGA)、多媒体处理器、矢量处理器等,协调地执行程序,以实现高性能和降低功耗。

异构计算架构已经存在了很多年,例如:

(1) 在超级计算机领域,将 CPU 与矢量处理器结合起来的矢量超级计算机最早于 20 世纪 70 年代后半叶商业化。

(2) 在消费电子产品中,异构计算被广泛用于各种设备。视频游戏机:例如 2005 年开发的用于视频游戏控制台的 LSI 嵌入了一个 CPU 核和八个单指令多数据(SIMD)处理器核,以执行需要实时计算大量三维图形数据的游戏软件。智能手机:例如 2016 年开发的智能手机系统芯片(SoC)包括四个 CPU 核、一个 GPU 核和一个 DSP 核,以灵活支持各种移动应用。

(3) 异构计算也在向云计算扩展。微软公司一直在使用 FPGA 加速搜索引擎和云服务的机器学习,从而降低功耗。谷歌公司一直在使用 GPU 加速机器学习,还开发了专门用于机器学习加速的张量处理单元(TPU)。此外,亚马逊公司的云服务提供了异构计算平台,其中 GPU 和 FPGA 被用作加速器。

近年,GPU 和 FPGA 作为异构计算的加速器受到了广泛关注。由于半导体小型化的好处,可以集成在芯片上的计算单元的数量持续增加,计算性能显著提高。2017 年,

GPU 和 FPGA 的单精度浮点性能已达到每秒 10 万亿次浮点运算（TFLOPS），远远超过 CPU 的计算性能。由于 GPU 和 FPGA 具有很强的计算能力，它们已在许多领域用作加速器，包括图像/信号处理、密码处理、生物信息学、CAD/CAE/CAM 应用、流体动力学、金融工程和机器学习等。

GPU 是一种处理器类型的加速器，它将许多计算单元集成到一个芯片上。GPU 软件设计人员需要开发一个并行程序，该程序以 SIMD 方式并行运行在嵌入式计算单元上，目前已经为这种并行编程开发了 CUDA 和开放计算语言（OpenCL）等软件开发环境，使软件工程师可以更容易地开发 GPU 程序。

与 GPU 不同，FPGA 的硬件结构可以根据需求进行定制，可以自由配置其内部逻辑，因此 FPGA 加速器可以采用任何类型的架构，如数据流、脉动阵列（systolic array）和 SIMD/多指令多数据（MIMD）处理器。通过为算法或处理选择最佳架构，FPGA 加速器可以实现比 GPU 更高的总体性能和能效比。然而，在实现一个 FPGA 加速器时，首先需要在 FPGA 上构建硬件算法的目标体系结构，这种硬件设计过程有时是软件工程师放弃使用 FPGA 加速器的一个因素。

1.2　OpenCL

OpenCL 的英文全称是 Open Computing Language，即开放计算语言，是第一个面向异构系统并行编程的开放式、免费标准，也是一个统一的编程环境。OpenCL 便于软件开发人员为高性能计算服务器、桌面计算系统、手持设备编写高效便捷的代码，并广泛适用于多核处理器（CPU）、图形处理器（GPU）、Cell 类型架构以及数字信号处理器（DSP）等其他并行处理器，在游戏、娱乐、科研、医疗等各种领域都有广阔的发展前景。

OpenCL 最初由苹果公司开发，拥有其商标权，并在与 AMD、IBM、英特尔和 NVIDIA 技术团队的合作之下初步完善。随后，苹果公司将这一草案提交至 Khronos Group。

在 2008 年 6 月的 WWDC 大会上，苹果公司提出了 OpenCL 规范，旨在提供一个通用的开放 API，在此基础上开发 GPU 通用计算软件。随后，Khronos Group 宣布成立 GPU 通用计算开放行业标准工作组，以苹果公司的提案为基础创立 OpenCL 行业规范。2008 年 11 月 18 日，该工作组发布了 OpenCL 1.0 规范的技术细节。2010 年 6 月 14 日，OpenCL 1.1 发布。2011 年 11 月 15 日，OpenCL 1.2 发布。2013 年 11 月 19 日，OpenCL 2.0 发布。图 1-1 为 Khronos Group 成员 LOGO。

OpenCL 工作组的成员包括 3Dlabs、AMD、苹果、ARM、Codeplay、爱立信、飞思卡尔、华为、HSA 基金会、Graphic Remedy、IBM、Imagination Technologies、Intel、诺基亚、NVIDIA、摩托罗拉、QNX、高通、三星、Seaweed、德州仪器、布里斯托尔大学、瑞典 Ume 大学。

图 1-1　Khronos Group 成员 LOGO

OpenCL 也可以认为是一种异构计算架构,它包括面向多平台的通用编程模型、宿主机 API 和 kernel 语言、基于 C/C++ 的编程语言,可以结合硬件加速器实现性能提升。

若多个计算平台是异构的,则需要不同的开发语言,这增加了开发难度。如图 1-2(a) 所示,系统存在 4 类异构计算系统:CPU、DSP、GPU 和 FPGA,每一个开发平台采用的开发软件都不同。为了完成整个系统的开发,需要掌握 4 种开发软件及相关语言或需要 4 类具有相关知识技能的人员共同完成。与之形成鲜明对比的是,若使用 OpenCL,则只需要掌握 OpenCL 语言及相关开发软件即可应对 4 类异构计算平台,如图 1-2(b) 所示。

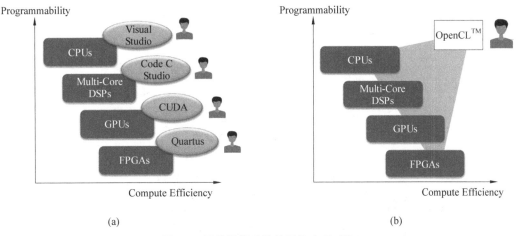

图 1-2 异构计算系统的开发方式对比

采用一种语言实现异构平台的开发不仅可以使开发者优化系统性能,而且可以使异构计算系统自动无缝连接,这也是 OpenCL 实现异构计算的优势所在。

1.3 FPGA

FPGA(Field-Programmable Gate Array)是现场可编程门阵列的简称。FPGA 是一种可编程硬件,属于半定制器件,可以在制造后根据应用需要重新配置。如图 1-3 所示,FPGA 的基本结构包含 3 部分:可编程逻辑(configurable logic block,CLB)、可编程互连(programable interconect,PI)和可编程 IO(programable input/output,PIO),为了灵活布线,在可编程连线的某些位置放置了可编程开关矩阵(programable switch maxtrix,PSM)。

由于集成工艺的不断发展,现代 FPGA 还包含可配置的内存模块和 DSP(一般由专用乘法器构建)等,如图 1-4 所示。

开发人员可以通过 EDA 软件并利用片上的逻辑门、DSP 和内存实现任意电路。因此,同样可以在 FPGA 上实现多个处理器或加速器以进行不同的计算。FPGA 已经在信号处理、高性能计算、机器学习等许多领域得到应用。图 1-5、图 1-6、图 1-7 给出了在实现不同应用时 FPGA 计算平台与其他硬件平台的性能对比。

图 1-5 为实现 AES 加密算法时,采用 E5503 Xeon Processor(CPU)、AMD Radeon

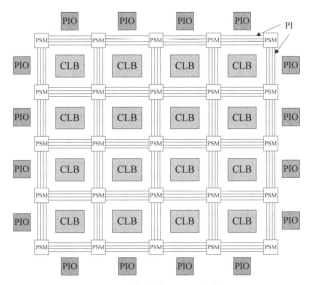

图 1-3　典型的 FPGA 结构

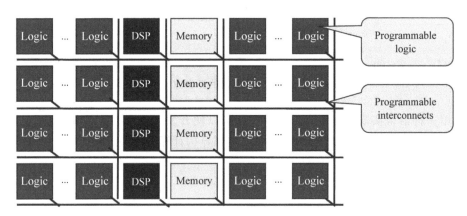

图 1-4　集成特殊功能单元的 FPGA 结构

AES Encryption	
Platform	Throughput (GB/s)
E5503 Xeon Processor	0.01 (single core)
AMD Radeon HD 7970	0.33
PCIe385 A7 Accelerator (Stratix V FPGA)	5.20

图 1-5　AES 加密算法实现

HD 7970(GPU)与 PCIe385 A7 Accelerator(FPGA)的吞吐率的对比。

图 1-6 为实现多资产障碍期权定价的 Monte Carlo 仿真时，采用 W3690 Xeon Processor（CPU）、nVidia Tesla C2075（GPU）与 PCIe385 D8 Accelerator（FPGA）的能耗、性能及能效比的对比。

图 1-7 为实现文献筛选时，采用 W3690 Xeon Processor（CPU）、nVidia Tesla C2075

Multi-Asset Barrier Option Pricing via Monte Carlo Simulation			
Platform	Power (W)	Performance (Msims/s)	Msims/W
W3690 Xeon Processor	130	32	0.25
nVidia Tesla C2075	225	63	0.28
PCIe385 D5 Accelerator (Stratix V FPGA)	23	170	7.4

图 1-6 多资产障碍期权定价的 Monte Carlo 仿真

Document Filtering (4K Points)			
Platform	Power (W)	Performance (MTs)	MTs/W
W3690 Xeon Processor	130	2070	15.92
nVidia Tesla C2075	215	3240	15.07
PCIe385 A7 Accelerator (Stratix V FPGA)	25	3602	144.08

图 1-7 文献筛选

(GPU)与 PCIe385 D8 Accelerator(FPGA)的能耗、性能及能效比的对比。

1.4 FPGA+CPU 异构计算系统

FPGA 现在被用于开发各种领域的定制加速器,涵盖从低功耗嵌入式应用到高性能计算的各个方面。由于半导体小型化技术的优势,FPGA 的规模不断扩大,并包含各种硬件资源,如逻辑块、DSP 块、内存块和 CPU 核心。此外,为了开发大规模的系统,FPGA 通过外围组件互连 Express(PCIe)等高速接口连接到 CPU,作为硬件加速器,可以形成高性能计算和大数据处理的集群。

图 1-8 显示了两种基于 CPU 和 FPGA 的异构计算系统。图 1-8(a)显示了一个 SOC (片上系统)类型的计算系统,其中 CPU 和 FPGA 集成在同一封装或同一芯片上。CPU 和 FPGA 之间的数据传输是通过内部(片上)总线完成的。这种类型的系统通常用于低功耗嵌入式处理。图 1-8(b)显示了另一种带有 CPU 和 FPGA 的计算系统。在该系统中,FPGA 通过外部总线(如 PCI Express(PCIe))连接到 CPU,CPU 和 FPGA 之间的数据传输通过外部总线完成,这种类型的系统通常用于高性能计算。本书主要介绍图 1-8 (a)所描述的通过内部(片上)总线连接 FPGA 的异构计算系统。

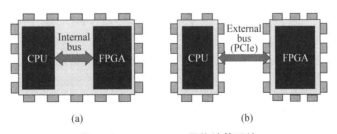

图 1-8 FPGA+CPU 异构计算系统

不同的 FPGA 开发板具有不同的输入/输出(I/O)资源,如外部存储器、PCI Express、USB 等。此外,对于不同用途,如低功耗需求和高性能计算两种不同的应用场景,FPGA 需要选择合适的型号,不同型号 FPGA 的逻辑和内存资源明显不同。因此,在 I/O、逻辑和内存资源未知的情况下,通常很难在不同的 FPGA 开发板上使用相同的代码。

为了在不同的开发板上运行相同的代码,提高 OpenCL 的可移植性,图 1-9 显示了解决这个问题的方法。OpenCL 使用一个包含逻辑和内存等硬件信息的板支持包(BSP)将 OpenCL 内核代码独立于硬件信息,如图 1-9 中的 BSP 包含 DDR3、DMA、PCIe 及 IO 管脚等硬件信息。在编译过程中,内核 kernel 与 BSP 合并,kernel 与 BSP 之间通过编译生成的连线逻辑实现信息交互。因此,内核可以通过 BSP 访问 I/O。由于 BSP 负责 I/O 和 FPGA 资源,因此可以在任何支持 OpenCL 的 FPGA 开发板中使用相同的内核代码,前提是针对每一款 FPGA 开发板,需要提供对应的 BSP。

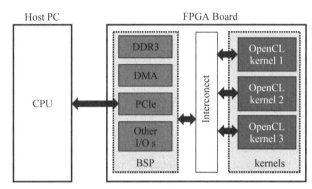

图 1-9 BSP 实现 OpenCL 设计兼容多种硬件平台

为了使用 OpenCL 内核,用户必须使用目标 FPGA 开始板的 BSP 信息对其进行编译。当离线编译器版本更改时,某些 BSP 可能无法与较新版本一起使用。在这种情况下,FPGA 开发平台供应商需要提供新的 BSP 版本。用户也可以编写自己的 BSP。然而,设计自己的 BSP 是一项困难的任务,它需要用户对 FPGA 有广泛的了解。

1.5 HDL 和 OpenCL

与专用集成电路(ASIC)相比,FPGA 由于其硬件可编程的特点,是一种灵活、快速、易升级的设计载体。硬件描述语言,如 Verilog HDL 和 VHDL 是 FPGA 系统开发的主要设计输入方式。在加速领域,为了设计高性能加速器,程序员必须对硬件设计有丰富的知识和经验。此外,基于 HDL 的设计需要周期较长的模拟和调试。因此,加速器的设计时间非常长。

在复杂的大型系统设计中,使用硬件描述语言(HDL)的传统设计方法效率较低,因为对于设计人员来说,为这种基于 FPGA 的大型复杂系统开发逻辑功能和控制状态的详细信息更加困难。

与 HDL 方法相对应,基于 C 语言的设计方法比基于 HDL 的方法更有效,因为 C 语言允许开发人员定义高级行为描述。然而,即使设计人员使用基于 C 语言的设计工具设计一个完整的系统也存在困难。基于 C 语言的设计工具只能在 FPGA 中生成数据路径，不支持 FPGA 和外部存储器设备之间的接口，也不支持 FPGA 和主机 CPU 之间的接口，设计人员仍然需要使用 HDL 设计接口电路。

为了解决这些问题，FPGA 设计中引入了一个基于 C 语言的 OpenCL 开发环境。OpenCL 允许设计者描述整个计算：主机上的计算、主机和加速器之间的数据传输以及加速器上的计算。

1.5.1 OpenCL 的优点

1. 用 C 语言设计

在基于 OpenCL 的设计中，设计人员可以使用类 C 语言的 OpenCL 代码设计一个 FPGA 加速器，可以大幅缩短设计时间。

基于 HDL 的设计在编码、模拟和调试方面需要更长的时间。

2. 支持 I/O

对于 FPGA，OpenCL C 附带许多常见的 I/O 控制器，如内存控制器、PCIe 控制器、DMA 控制器等，它还包含 PCIe 设备驱动程序和应用程序编程接口（API），用于控制 FPGA 和传输数据。

HDL 设计人员需要设计所有这些控制器，以及设备驱动程序和 API。不仅需要很长的设计时间和大量的工作，而且还需要全面了解 I/O 控制器的工作原理和实现方法。

3. 兼容并可在不同类型的 FPGA 开发板上重复使用

为一块 FPGA 开发板设计的 OpenCL 内核代码也适用于其他类型的 FPGA 开发板，这种可移植性是通过使用特定类型开发板的 BSP 重新编译代码完成的。

由于外部存储器（DDR3、DDR4 等）、存储器容量、存储器模块数量、逻辑资源、DSP 资源、网络 I/O 等方面的差异，一块开发板的 HDL 代码可能无法在不同类型的开发板上工作。因此，当使用不同类型的开发板时，通常需要进行较大的改动，甚至需要重新设计。

4. 易于调试

OpenCL 代码的功能行为可以在 CPU 上模拟。可以在内核中使用 printf 命令轻松收集中间结果。用户还可以编译内核进行分析，并在运行时收集内存访问等信息。

基于 HDL 的设计采用了逻辑级仿真。逻辑级仿真比基于 OpenCL 设计的功能级仿真慢很多。此外，为了在运行时收集中间结果，需要设计在 FPGA 内部进行调试的电路。调试电路的实现需要大量的时间和知识。

1.5.2 OpenCL 的缺点

1. 架构对设计人员是隐藏的

OpenCL 设计人员不直接设计体系结构，只编写 OpenCL 代码，离线编译器自动将其转换为 HDL 设计。因此，设计人员对生成的体系结构没有完整的了解，只能猜测得到的架构，并通过观察加速器的行为确认这些猜测。因此，为了获得更好的体系结构，设计人员需要知道离线编译器如何将 OpenCL 代码转换为硬件逻辑；否则，提高性能并不容易。

2. 无法设计指定的时钟频率

时钟频率由离线编译器自动确定。因此，编程器不能为了低功耗等目的降低时钟频率。

3. 难以控制资源利用

离线编译器通常以获得最大的性能为目标对内核进行编译。因此，即使某些内核不需要高性能，程序员也无法为低功耗等目的节省资源。

虽然编写 OpenCL 代码很容易，但要将其调整到最佳性能却很困难。因此，OpenCL 代码的性能调整和自动优化将是一个重要问题。由于 FPGA 集成度越来越高，人们很难在有限的设计时间内有效地利用所有 FPGA 资源设计最优的代码。因此，如果把设计时间作为一个首选衡量指标，那么基于 OpenCL 的设计的性能将超过基于 HDL 的设计。即使在当前，基于 OpenCL 的设计的性能也不远落后于基于 HDL 的设计。

图 1-10 给出了在 GZIP 文件压缩应用中，OpenCL 与 RTL 不同参考指标的对比。对比可以发现 OpenCL 实现方式不仅对设计人员的资质要求较低，而且需要更少的时间。缺点是具有略低的吞吐率和使用较多的逻辑资源。

Metric	OpenCL™	RTL
Experience	Summer Intern	FPGA Eng
Design Time	1 Month	3 Months
Throughput	2.7 GB/s	3.0 GB/s
Area	12% larger	

图 1-10 OpenCL 与 RTL 实现 GZIP 文件压缩功能的指标对比

注：Summer Intern 为暑期实习生。

1.6 人工神经网络

1.6.1 人工神经网络的基本概念

人工神经网络（artificial neural network，ANN）是 20 世纪 80 年代后人工智能领域兴起的研究热点。人工神经网络是一种非程序化、适应性、大脑风格的信息处理系统，其本质是通过网络的变换和动力学行为得到一种并行分布式的信息处理功能，并在不同程度和层次上模仿人脑神经系统的信息处理功能，是涉及神经科学、思维科学、人工智能、计算机科学等多个领域的交叉学科。

图 1-11 中包含两个人脑神经元。树突是接收从其他神经元传入的信息的入口，轴突为神经元的输出通道，作用是将细胞体发出的神经冲动传递给另一个或多个神经元。突触是指一个神经元的输出传到另一个神经元的相互接触的结构。

人工神经网络从信息处理角度对人脑神经元（图 1-11）网络进行抽象，从而建立某种简单模型（图 1-12），按不同的连接方式组成不同的网络（图 1-13）。在工程与学术界也常直接简称其为神经网络或类神经网络。神经网络通常可以看作是对自然界某种算法或者

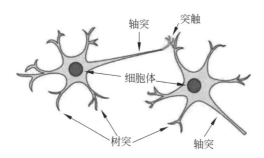

图 1-11 人脑神经元结构

函数的逼近,或是对一种逻辑策略的表达。

图 1-12(a)为单个神经元的数学模型。在该模型中,输入 x 代表输入的信息,权值 w 代表神经元的输入突触,wx 则对应生物神经元的树突,加权求和代表对输入信息的处理,f 为激活函数对应生物神经元的轴突。图 1-12(b)是更常用的神经元模型的画法。

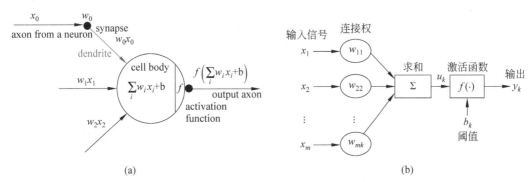

图 1-12 神经元模型

由以上分析可知,神经网络可以看作是一种运算模型,由大量的节点(或称神经元)相互连接构成。每个节点代表一种特定的输出函数,称为激活函数(activation function)。每两个节点间的连接都代表一个对于通过该连接信号的加权值,称为权重或权值,这相当于人工神经网络的记忆。网络的输出取决于网络的连接方式、权重值和激励函数。

人工神经网络中,神经元处理单元可处理不同的对象,例如特征、字母、概念或者一些有意义的抽象模式。网络中,处理单元的类型分为三个层次:输入层、输出层和隐形层,如图 1-13 所示。输入层接收外部世界的信号与数据;输出层实现系统处理结果的输出;隐形层处在输入和输出层之间,不能由系统外部观察。神经元间的连接权值反映了单元间的连接强度,信息的表示和处理体现在网络处理单元的连接关系中。一个神经网络可以含有多个隐形层。

1.6.2 人工神经网络的基本特征

人工神经网络是并行分布式系统,采用了与传统人工智能和信息处理技术完全不同的机理,克服了传统的基于逻辑符号的人工智能在处理直觉、非结构化信息方面的缺陷,具有自适应、自组织和实时学习的特点。

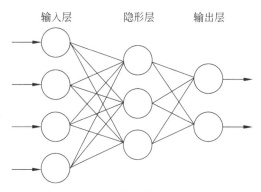

图 1-13　人工神经网路

人工神经网络是由大量处理单元互联组成的非线性、自适应信息处理系统,它是在现代神经科学研究成果的基础上提出的,试图通过模拟大脑神经网络处理和记忆信息的方式进行信息处理。人工神经网络具有四个基本特征。

(1) 非线性。非线性关系是自然界的普遍特性。大脑的智慧就是一种非线性现象。人工神经元处于激活或抑制这两种不同的状态,这种行为在数学上表现为一种非线性关系。具有阈值的神经元构成的网络具有更好的性能,可以提高容错性和存储容量。

(2) 非局限性。一个神经网络通常由多个神经元广泛连接而成。一个系统的整体行为不仅取决于单个神经元的特征,而且可能由单元之间的相互作用、相互连接所决定。通过单元之间的大量连接可以模拟大脑的非局限性。联想记忆是非局限性的典型例子。

(3) 非常定性。人工神经网络具有自适应、自组织、自学习能力。神经网络不仅可以处理各种变化的信息,而且在处理信息的同时,神经网络本身也在不断变化。通常采用迭代过程描述神经网络的演化过程。

(4) 非凸性。一个系统的演化方向在一定条件下取决于某个特定的状态函数。例如能量函数,它的极值对应于系统比较稳定的状态。非凸性是指这种函数有多个极值,故系统具有多个较稳定的平衡态,这将导致系统演化的多样性。

1.6.3　人工神经网络的应用

最近十多年来,人工神经网络的研究工作不断深入,已经取得了很大进展,其在模式识别、智能机器人、自动控制、预测估计、生物、医学、经济等领域已成功解决了许多现代计算机难以解决的实际问题,表现出了良好的智能特性。

1. 人工神经网络在信息领域中的应用

在许多问题中,信息来源既不完整,又包含假象,决策规则有时相互矛盾,有时无章可循,这给传统的信息处理方式带来了很大的困难,而神经网络却能很好地处理这些问题,并给出合理的识别与判断。

(1) 信息处理。

人工神经网络具有模仿或代替与人的思维有关的功能,可以实现自动诊断、问题求解,解决传统方法所不能或难以解决的问题。人工神经网络系统具有很高的容错性、鲁棒性及自组织性,即使连接线遭到很高程度的破坏,它仍能处在优化工作状态,这点在军事系统电子设备中得到了广泛应用。现有的智能信息系统包括智能仪器、自动跟踪监测仪器系统、自动控制制导系统、自动故障诊断和报警系统等。

(2) 模式识别。

模式识别是指对表征事物或现象的各种形式的信息进行处理和分析,从而对事物或现象进行描述、辨认、分类和解释的过程。该技术以贝叶斯概率论和香农的信息论为理论基础,对信息的处理过程更接近人类大脑的逻辑思维过程。现在有两种基本的模式识别方法,即统计模式识别方法和结构模式识别方法。人工神经网络是模式识别中的常用方法,近年来发展起来的人工神经网络模式的识别方法逐渐取代了传统的模式识别方法。经过多年的研究和发展,模式识别已成为当前比较先进的技术,被广泛应用到文字识别、语音识别、指纹识别、遥感图像识别、人脸识别、手写体字符识别、工业故障检测、精确制导等领域。

2. 人工神经网络在医学中的应用

由于人体和疾病的复杂性与不可预测性,以及在生物信号与信息的表现形式及变化规律上,对其进行检测与信号表达,获取的数据及信息的分析、决策等诸多方面都存在非常复杂的非线性联系,因此非常适合人工神经网络的应用。目前的研究几乎涉及从基础医学到临床医学的各个方面,主要应用在生物信号的检测与自动分析、医学专家系统等领域。

(1) 生物信号的检测与自动分析。

大部分医学检测设备都是以连续波形的方式输出数据,这些波形是诊断的依据。人工神经网络是由大量的神经元节点连接而成的自适应动力学系统,具有巨量并行性、分布式存储、自适应学习的自组织等功能,可以解决生物医学信号分析处理中常规法难以解决或无法解决的问题。神经网络在生物医学信号检测与处理中的应用主要集中在脑电信号的分析,听觉诱发电位信号的提取,肌电和胃肠电等信号的识别,心电信号的压缩,医学图像的识别和处理等方面。

(2) 医学专家系统。

传统的专家系统是把专家的经验和知识以规则的形式存储在计算机中,建立知识库,用逻辑推理的方式进行医疗诊断。但在实际应用中,随着数据库规模的增大,这样做将导致知识"爆炸",在知识获取途径中也存在瓶颈,致使工作效率很低。

以非线性并行处理为基础的神经网络为专家系统的研究指明了新的发展方向,解决了专家系统的以上问题,并提高了知识的推理、自组织、自学习能力,从而使神经网络在医学专家系统中得到了广泛的应用和发展。

在麻醉与危重医学等相关领域的研究中涉及多生理变量的分析与预测,在临床数据中存在着一些尚未发现或无确切证据的关系与现象,信号的处理、干扰信号的自动区分检测、各种临床状况的预测等都可以应用到人工神经网络技术。

3. 人工神经网络在经济领域的应用

(1) 市场价格预测。

对商品价格变动的分析可归结为对影响市场供求关系的诸多因素的综合分析。传统的统计经济学方法因其固有的局限性,难以对价格变动做出科学的预测,而人工神经网络容易处理不完整的、模糊不确定或规律性不明显的数据,所以用人工神经网络进行价格预测有着传统方法无法比拟的优势。从市场价格的确定机制出发,依据影响商品价格的家庭户数、人均可支配收入、贷款利率、城市化水平等复杂多变的因素,建立较为准确可靠的模型,该模型可以对商品价格的变动趋势进行科学预测,并得到准确客观的评价结果。

(2) 风险评估。

风险是指在从事某项特定活动的过程中,因其存在的不确定性而产生的经济或财务的损失、自然破坏或损伤的可能性。防范风险的最佳办法就是事先对风险做出科学的预测和评估。应用人工神经网络的预测思想是根据具体现实的风险来源构造出适合实际情况的信用风险模型的结构和算法,得到风险评价系数,然后确定实际问题的解决方案。利用该模型进行实证分析能够弥补主观评估的不足,可以取得令人满意的效果。

4. 人工神经网络在控制领域中的应用

人工神经网络由于其独特的模型结构和固有的非线性模拟能力,以及高度的自适应和容错特性等突出特征,在控制系统中获得了广泛应用。其在各类控制器框架结构的基础上加入了非线性自适应学习机制,从而使控制器具有更好的性能。基本的控制结构包括监督控制、直接逆模控制、模型参考控制、内模控制、预测控制、最优决策控制等。

5. 人工神经网络在交通领域的应用

交通运输同样是一个高度非线性问题,可获得的数据通常是大量的、复杂的,用神经网络处理相关问题有巨大的优越性,其应用范围涉及汽车驾驶员行为的模拟、参数估计、路面维护、车辆检测与分类、交通模式分析、货物运营管理、交通流量预测、运输策略与经济、交通环保、空中运输、船舶自动导航及船只辨认、地铁运营及交通控制等领域,并已经取得了很好的效果。

6. 人工神经网络在心理学领域的应用

从神经网络模型的形成开始其就与心理学就有着密不可分的联系。神经网络抽象于神经元的信息处理功能,神经网络的训练则反映了感觉、记忆、学习等认知过程。人们通过不断研究,调整人工神经网络的结构模型和学习规则,从不同角度探讨着神经网络的认知功能,为其在心理学的研究中奠定了坚实的基础。近年来,人工神经网络模型已成为探讨社会认知、记忆、学习等高级心理过程机制不可或缺的工具。人工神经网络模型还可以对脑损伤病人的认知缺陷进行研究,对传统的认知定位机制提出了挑战。

习题 1

1.1 异构计算系统有哪些特点?给出几个异构计算系统的例子。

1.2 什么是 OpenCL?OpenCL 在异构计算中有哪些优势?

1.3 什么是 FPGA？简述 FPGA 的基本结构。
1.4 FPGA＋CPU 构成的异构计算系统的类型有哪两类？各有什么特点？
1.5 OpenCL 如何解决不同 FPGA 异构平台上的移植问题？
1.6 基于 OpenCL 和 HDL 的 FPGA 开发方式各有什么优缺点？
1.7 人工神经网络模型中包含哪些基本元素？
1.8 简述人工神经网络中的神经元模型。
1.9 人工神经网络有哪些特征？
1.10 人工神经网络的应用领域有哪些？

第 2 章

TensorFlow 基础知识及运行环境搭建

本章首先介绍 TensorFlow 编程模式的基本操作，然后介绍在 Ubuntu 系统下搭建 TensorFlow 运行环境的方法，最后介绍在搭建好的环境下运行 TensorFlow 例程的方法。

2.1 TensorFlow 简介

TensorFlow 是一个通过计算图的形式表述计算的编程系统，它灵活的架构让用户可以在多种平台上展开计算，例如台式计算机中的一个或多个 CPU（或 GPU）、服务器、移动设备等。TensorFlow 最初由谷歌公司大脑小组（隶属于谷歌公司机器智能研究机构）的研究员和工程师开发出来，用于机器学习和深度神经网络方面的研究，但这个系统的通用性使其也可广泛用于其他计算领域。

TensorFlow 的名称可以分解为 Tensor 和 Flow 两个部分。Tensor 指张量，在 TensorFlow 中可以简单理解为多维数组，表示数据的结构。Flow 指数据的流动，表示张量之间通过计算相互转换的过程。TensorFlow 中的每一个计算都表示为计算图中的节点（nodes），而节点之间的边线（edges）描述了计算之间的依赖关系，即张量在节点之间的流动路径。由一条边线连接的两个节点可以理解为：前级节点的计算结果保存在一个张量中，该张量作为后级节点的输入张量。

2.2 TensorFlow 两步编程模式

TensorFlow 程序一般分为两个阶段：定义计算图和运行计算图。在第一个阶段需要定义计算图的结构，包括所有的计算及张量。第二个阶段需要根据计算图的结构执行计算。

1. 定义计算图

在定义计算图时，只要确定了节点和边线，就可以完成计算图的定义。计算图结构的示意如图 2-1 所示。图中含有两个基本元素：节点和边线。节点在计算图中表示数学运算(operation)，边线则表示在节点间相互联系的多维数据数组，即张量(tensor)。图 2-2 列举了一个计算图例。

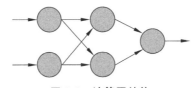

图 2-1　计算图结构

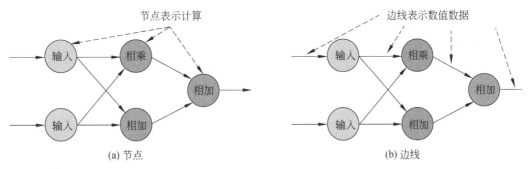

图 2-2　计算图例

2. 运行计算图

运行计算图是指输入数据在定义好的计算图中经过节点计算并通过边线流向输出的过程。图 2-3 描述了计算图运行的示意图。

图 2-3　计算图的运行

2.3　TensorFlow 两步编程模式实例

本教材只介绍 TensorFlow 的基本操作，详尽的语法知识请参考相关的 TensorFlow 及 Python 教程。

2.3.1　定义计算图的基本操作

1. 导入 TensorFlow

在 Python 中，一般会采用 import tensorflow as tf 的形式载入 TensorFlow，即用 tf 代替 tensorflow 作为模块的名称，使整个程序更加简洁。该语句自动创建默认图，默认图只提供了模型运算的默认位置，是一个空白图，程序根据之后的代码在该位置生成对应计算图，如图 2-4 所示。

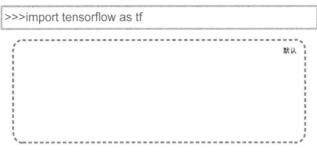

图 2-4　导入 TensorFlow

2. 创建常量数据

使用 API 函数 tf.constant()可以创建常量数据值,该函数返回一个与参数相同的常数,该函数也可以看作是生成一个常数的运算,运算的输出称为张量。如图 2-5 所示,a=tf.constant(3.0)定义了一个运算节点,该节点的输出值为 3.0。tf.constant 代表运算节点,a 代表运算的张量输出。

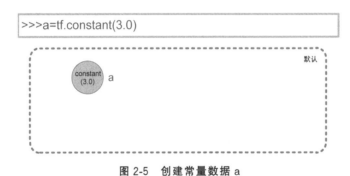

图 2-5　创建常量数据 a

tf.constant()除了可设置常数参数外,还可以设置一个名称参数。a=tf.consant(3.0,name="input1")语句生成了一个名称为 input1 的运算节点,该运算节点的张量输出为a,输出值为 3.0,如图 2-6 所示。

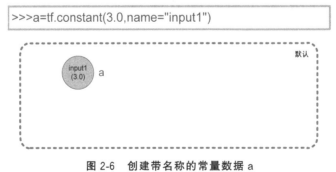

图 2-6　创建带名称的常量数据 a

我们还可以生成其他运算节点,b=tf.constant(2.0,name="input2")语句生成一个名称为 input2 的运算节点,该节点的张量输出为 b,输出值为 2.0,如图 2-7 所示。

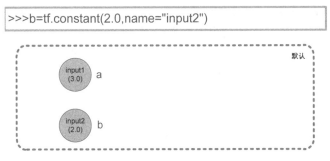

图 2-7 创建带名称的常量数据 b

3. 创建加法运算节点

图 2-7 中,变量 a 和 b 代表两个 constant operation 的张量输出。这两个张量可以作为下一个数学计算(节点)的输入。语句 c=tf.add(a,b,name="my_add_op")创建了一个名称为 my_add_op 的加法运算节点,输入张量为 a、b,输出张量为 c,如图 2-8 所示。

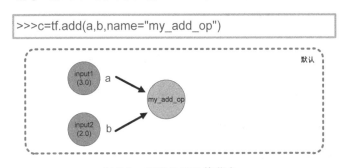

图 2-8 创建加法运算节点 c

4. 创建乘法运算节点

语句 d=tf.multiply(a,c,name="my_mul_op")创建了一个名称为 my_mul_op 的乘法运算节点,输入张量为 a、c,输出张量为 d,如图 2-9 所示。

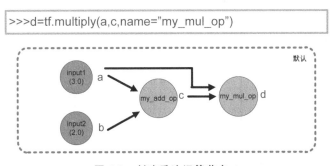

图 2-9 创建乘法运算节点 d

5. 定义运行时可变的输入变量

在图 2-9 中，input1 和 input2 的运算结果为固定的常数，在运行计算图的过程中不会发生改变。若希望输入值在运行过程中发生改变，则可使用 tf.placeholder()函数定义一个输入变量，该输入变量可在图运行的过程中通过读取字典得到不同的输出值。

执行语句 a＝ tf.placeholder（tf.float32，name＝"input1"）、b＝ tf.placeholder（tf.float32，name＝"input2"）分别代替 a＝tf.constant（3.0，name＝"input1"）、b＝tf.constant（2.0，name＝"input1"），计算图如图 2-10 所示。对比图 2-9 和图 2-10，节点 input1 和 input2 的值不再固定为 3.0 和 2.0。图 2-10 中的 input1 和 input2 的值在计算图的运行过程中确定。

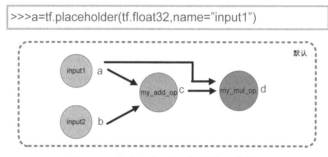

图 2-10　定义运行时可变的输入变量 a

tf.placeholder()中的参数 tf.float32 表示输入值的数据类型为 32 位 float 类型。

2.3.2　运行计算图的基本操作

1. 定义 Session 执行对象

TensorFlow 使用 tf.Session()函数实现计算图的执行。每个 Session 对象专用于执行一个计算图。如图 2-11 所示，语句 sess＝tf.Session()定义了名为 sess 的执行对象。Sess 拥有并管理着 TensorFlow 程序运行的所有资源。所有计算完成之后，需要通过关闭 sess 帮助系统回收资源，否则会出现资源泄露的问题。

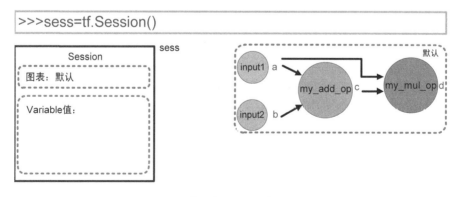

图 2-11　定义名为 sess 的执行对象

默认情况下,Session 使用当前默认的计算图,如图 2-12 所示。

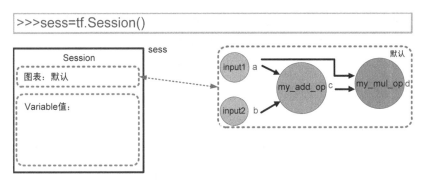

图 2-12　Session 使用的计算图

2. 定义填充 placeholder 张量数据的字典

使用 tf.placeholder()函数定义的运算节点只定义了一个占位符,没有赋予数值,因此在运行图时需要填充数据,如图 2-13 所示。

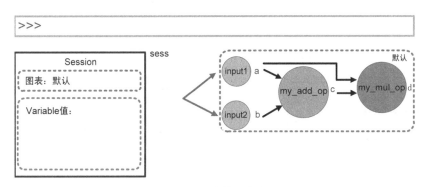

图 2-13　未填充数据

可以定义一个数值字典,实现输入变量的填充。语句 feed_dict＝{a:3.0,b:2.0}定义了一个数值字典,a 取值为 3.0,b 取值为 2.0,如图 2-14 所示。

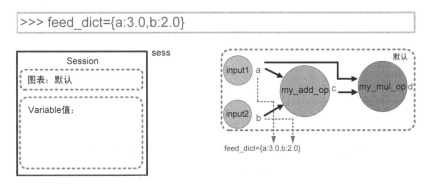

图 2-14　使用 feed_dict 填充数据

3. Session 对象的执行

为运行计算图,需要执行之前定义的 Session 对象 sess。run(fetches,feed_dict)函数可以实现 Session 对象的执行,fetches 参数指定输出张量,feed_dict 指定数据来源。

语句 out=sess.run(d,feed_dict=feed_dict)声明了输出张量为 d,数据来源为 feed_dict,如图 2-15 和图 2-16 所示。

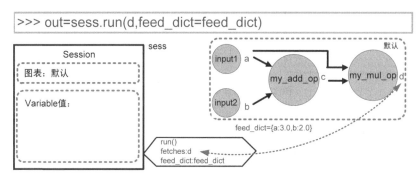

图 2-15 sess 运行的输出张量为 d

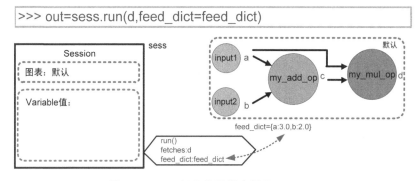

图 2-16 sess 运行的数据来源为 feed_dict

在语句 out=sess.run(d,feed_dict=feed_dict)中,feed_dict=feed_dict 参数等号右侧为用户定义的字典名称,可以更换为其他标识符。

4. 执行结果

语句 out=sess.run(d,feed_dict=feed_dict)执行结束后,可用 print(out)语句显示执行结果,如图 2-17 所示。

5. Variable()函数

为了提供计算图的运行信息或者保存训练权值数据,需要使用 Variable()函数。语句 store=tf.Variable(0.0,name="var")定义了一个名称为 var 的运算节点,取值为 0.0,张量输出为 store,该张量输出具有变量特性,如图 2-18 所示。

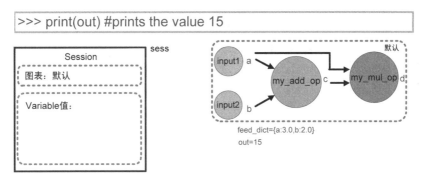

图 2-17　打印输出结果

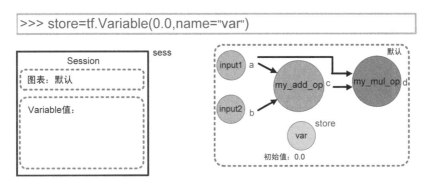

图 2-18　定义名称为 var 的运算节点 store

语句 inc＝store.assign(store＋1)定义了对变量 store 进行"加 1"的运算,输入张量为 store,输出张量为 inc,如图 2-19 所示。

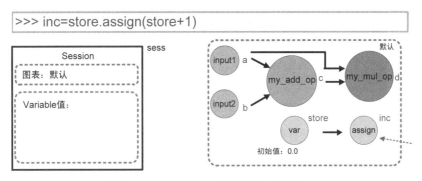

图 2-19　对变量 store 进行"加 1"运算 inc

注意：图 2-19 中 Session 中的 Variable 值为空白,若 Session 使用 Variable,则需要进行初始化,语句 init＝tf.global_variable_initializer()定义了对所有变量进行初始化的运算,如图 2-20 所示。

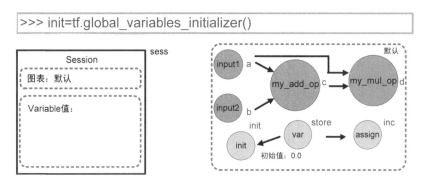

图 2-20 定义对所有变量进行初始化参数 init

语句 sess.run(init) 实现了对计算图中变量的初始化，如图 2-21 所示。init 读取 var 的初始值，并将其分配至 sess 中存储。

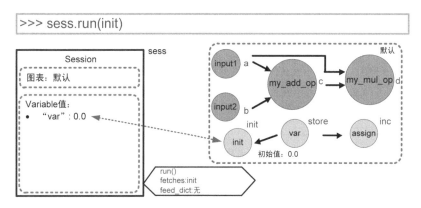

图 2-21 运行初始化参数 init

语句 sess.run(inc) 实现了对 var 的 inc＝store.assign(store＋1) 运算，如图 2-22 所示。执行 sess.run(inc) 语句后，var 的结果由 0.0 变为 1.0，如图 2-23 所示。

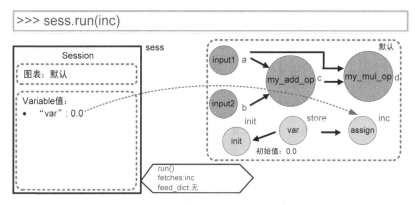

图 2-22 运行 sess.run(inc) 之前

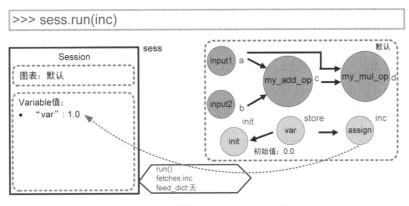

图 2-23 运行 sess.run(inc)之后

2.4 TensorFlow 环境搭建

本教材中 Ubuntu 命令格式的说明如下。
(1) root@用户名-计算机名:/路径名♯　命令。
(2) 用 $ 代指 root@用户名-计算机名:/路径名♯。

2.4.1 软件安装

1. 安装方式

基于 VirtualEnv 方式进行安装。

2. 安装步骤及命令

(1) 安装 VirtualEnv,运行终端命令:

```
$ sudo apt-get install python-pip python-dev python-virtualenv
```

(2) 建立一个全新的 virtualenv 环境,将环境建在/home/tensorflow 目录下,运行终端命令:

```
$ virtualenv --system-site-packages /home/tensorflow
```

执行上面的命令需要等待一段时间,当终端显示"done"时,上面的指令执行完成。

(3) 切换到/home/tensorflow 路径下,运行终端命令:

```
$ cd /home/tensorflow
```

(4) 激活 virtualenv,运行终端命令:

```
$ source bin/activate
```

如果安装成功,则在命令行前导字段会出现(tensorflow)。

(5) 在 virtualenv 环境下安装 TensorFlow,本教材使用的 TensorFlow 版本为 1.4.0。下载网址为 https://pypi.python.org/pypi/tensorflow/1.4.0。

打开下载网址后,选择 Navigation 项中的 Download files 标签,右侧将显示可以下载的 TensorFlow 版本,本教材选择 tensorflow-1.4.0-cp27-cp27mu-manylinux1_x86_64.whl,如图 2-24 所示。

图 2-24 下载界面

下载文件的路径默认为用户主文件夹的"下载"目录,如图 2-25 所示。

图 2-25 默认下载路径

(6) 修改/home/tensorflow 文件夹的访问属性,运行终端命令:

```
(tensorflow)$ chmod -R 777 /home/tensorflow
```

上述命令可以将/home/tensorflow 文件夹的访问权限修改为可写。

(7) 在 Ubuntu 系统中采用复制-粘贴的方法将"下载"文件中的 tensorflow-1.4.0-cp27-cp27mu-manylinux1_x86_64.whl 复制到/home/tensorflow 文件夹下。

(8) 安装 TensorFlow 压缩包。

```
(tensorflow)$ pip install --upgrade
/home/tensorflow/tensorflow-1.4.0-cp27-cp27mu-manylinux1_x86_64.whl
```

作者系统运行的安装命令界面如下：

```
(tensorflow) root@ubuntu602-System-Product-Name:/home/tensorflow# pip install --upgrade /home/tensorflow/tensorflow-1.4.0-cp27-cp27mu-manylinux1_x86_64.whl
```

（9）查看 TensorFlow 的版本及安装路径，可在终端输入如下查询命令：

```
(tensorflow)$python
import tensorflow as tf
tf.__version__
tf.__path__
```

tf._ _version_ _为查询 TensorFlow 版本的命令。

tf._ _path_ _为查询 TensorFlow 安装路径的命令。

作者系统运行的命令界面如下：

```
(tensorflow) root@ubuntu602-System-Product-Name:/home/tensorflow# python
Python 2.7.12 (default, Nov 12 2018, 14:36:49)
[GCC 5.4.0 20160609] on linux2
Type "help", "copyright", "credits" or "license" for more information.
>>> import tensorflow as tf
>>> tf.__version__
'1.4.0'
>>> tf.__path__
['/home/tensorflow/local/lib/python2.7/site-packages/tensorflow']
>>> exit()
(tensorflow) root@ubuntu602-System-Product-Name:/home/tensorflow#
```

exit()为退出 Python 编辑环境的命令。

提示：可以使用 virtualenv 在不同目录下安装不同版本的 TensorFlow，运行时激活对应的版本即可。可实现不同版本 TensorFlow 之间的切换。

2.4.2　TensorFlow 软件运行

每次使用 TensorFlow 都需要运行以下终端命令，进入 TensorFlow 运行环境。

```
cd /home/tensorflow/
source bin/activate
```

或直接运行命令：

```
source /home/tensorflow/bin/active
```

作者系统运行的命令界面如下：

```
root@ubuntu602-System-Product-Name:/home/ubuntu602# cd /home/tensorflow/
root@ubuntu602-System-Product-Name:/home/tensorflow# source bin/activate
(tensorflow) root@ubuntu602-System-Product-Name:/home/tensorflow#
```

2.4.3　计算图例程运行实例

本节将基础知识中给出的计算图例程在搭建好的平台中运行，一是验证搭建的 TensorFlow 运行环境是否成功，二是熟悉 TensorFlow 运行的基本步骤，三是熟悉

TensorFlow 的基础知识。

在运行下面的例程前,首先需要进入 TensorFlow 运行环境,请参考 2.4.2 节。

1. 例一:常数输入

(1) 在/home/tensorflow 目录下新建文件 design,运行终端命令:

```
mkdir /home/tensorflow/design
```

(2) 在 Ubuntu 用户界面下,在/home/tensorflow/design 目录下新建文本文档,并重命名为 tf_basic1.py。

(3) 打开新建文本文档 tf_basic1.py,输入以下代码并保存。

```python
import tensorflow as tf

a=tf.constant(3.0,name="input1")
b=tf.constant(2.0,name="input2")
c=tf.add(a,b,name="my_add_op")
d=tf.multiply(a,c,name="my_mul_op")

sess=tf.Session()
a_val=sess.run(a)
out=sess.run(d)
sess.close()

print(a)
print(a_val)
print(b)
print(c)
print(d)
print(out)
```

(4) 运行 tf_basic1.py 文档,运行终端命令:

```
python tf_basic1.py
```

可以看到输出结果如图 2-26 所示。

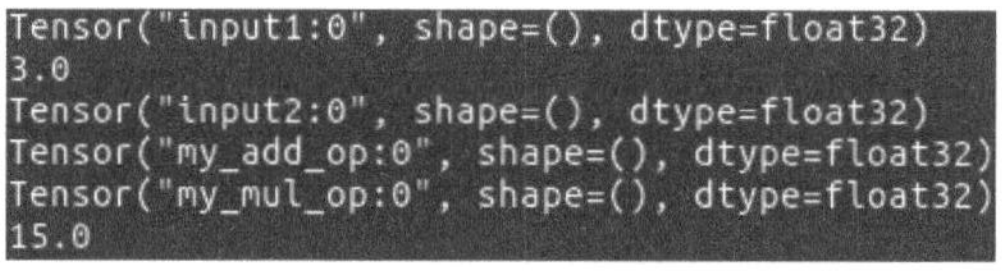

```
Tensor("input1:0", shape=(), dtype=float32)
3.0
Tensor("input2:0", shape=(), dtype=float32)
Tensor("my_add_op:0", shape=(), dtype=float32)
Tensor("my_mul_op:0", shape=(), dtype=float32)
15.0
```

图 2-26 例一运行结果

在 tf_basic1 例程中,a、b、c、d 分别为节点 input1、input2、my_add_op 和 my_mul_op 的输出张量。在 TensorFlow 中,张量代表一个结构,从 tf_basic1 例程的运行结果可以看

出,一个张量主要有三个属性:名称、维度和类型。

张量的第一个属性表示该张量是由哪个节点计算得到的,即节点的计算结果保存在张量中。例如,print(a)语句的执行结果为 Tensor(input1:0, shape=( ), dtype=float32),input1:0 表示张量 a 是节点 input1 的第一个输出。在这里,输出编号从 0 开始,因此 0 表示节点的第一个输出。

张量的第二个属性表示张量的维度信息。在 print(a)的执行结果中,shape=( )说明张量 a 为 0 维张量,即一个标量。若 shape=(3,),则说明对应的张量为 1 维张量,格式为 1×3。若 shape=(3,4,),则说明对应的张量为 2 维张量,格式为 3×4。

张量的第三个属性表示张量的类型。在 print(a)的执行结果中,dtype=float32 说明张量 a 的类型为 float32。TensorFlow 支持 14 种不同的类型,主要有实数(tf.float32、tf.float64)、整数(tf.int8、tf.int16、tf.int32、tf.int64、tf.uint8)、布尔型(tf.bool)和复数(tf.complex64、tf.complex128)。

要想得到张量代表的数值,就需要运行张量。例如,tf_basic1 中的 a_val= sess.run(a)表示对执行张量 a 得到张量 a 的运行结果 a_val,即其代表的数值。从图 2-26 可知,张量 a 的数值为 3.0。

sess=tf.Session()语句定义了一个会话(session),会话拥有并管理着 TensorFlow 程序运行的所有资源。所有计算完成之后,需要关闭会话以帮助系统回收资源,否则会出现资源泄露的问题。

TensorFlow 中使用会话的模式一般有两种,第一种需要明确调用会话生成函数和会话关闭函数。在该例程中,sess=tf.Session()为调用会话生成函数,sess.close()为调用会话关闭函数。采用这种模式可能在程序因为异常而退出时,会话关闭函数没有被执行而导致资源泄露。为解决异常退出时资源释放的问题,TensorFlow 采用 Python 的上下文管理器使用会话,即使用会话的第二种模式。第二种模式的格式采用 with tf.Session() as sess:语句,把所有的计算操作均放在该语句的内部。当上下文管理器退出时,程序会自动释放所有资源,这样既能解决因为异常退出而产生的资源释放问题,又能解决忘记调用会话关闭函数而导致的资源释放问题。

采用上下文管理器模式修改后的例一代码如下:

```
import tensorflow as tf

a=tf.constant(3.0,name="input1")
b=tf.constant(2.0,name="input2")
c=tf.add(a,b,name="my_add_op")
d=tf.multiply(a,c,name="my_mul_op")

with tf.Session() as sess:
    a_val=sess.run(a)
    out=sess.run(d)

print(a)
print(a_val)
```

```
print(b)
print(c)
print(d)
print(out)
```

需要注意的是,所有的计算都要放在 with 语句的内部,在编写代码时,需要将相关语句缩进相同的占位符,否则系统会提示 unexpected indent 的错误信息。

2. 例二:运行时可变输入

(1) 在 Ubuntu 用户界面下,在/home/tensorflow/design 目录下新建文本文档,并重命名为 tf_basic2.py。

(2) 打开新建文本文档 tf_basic2.py,输入以下代码并保存。

```
import tensorflow as tf

a=tf.placeholder(tf.float32,name="input1")
b=tf.placeholder(tf.float32,name="input2")
c=tf.add(a,b,name="my_add_op")
d=tf.multiply(a,c,name="my_mul_op")

sess=tf.Session()
feed_dict={a:3.0,b:2.0}
out=sess.run(d,feed_dict=feed_dict)

print(out)
sess.close()
```

(3) 运行 tf_basic2.py 文档,运行终端命令:

```
python tf_basic2.py
```

可以看到输出结果为 15.0。

下面为采用上下文管理器机制会话模式的代码示例。

```
import tensorflow as tf

a=tf.placeholder(tf.float32,name="input1")
b=tf.placeholder(tf.float32,name="input2")
c=tf.add(a,b,name="my_add_op")
d=tf.multiply(a,c,name="my_mul_op")
feed_dict={a:3.0,b:2.0}

with tf.Session() as sess:
    out=sess.run(d,feed_dict=feed_dict)

print(out)
```

3. 例三：变量使用

（1）在 Ubuntu 用户界面下，在 /home/tensorflow/design 目录下新建文本文档，并重命名为 tf_basic3.py。

（2）打开新建文本文档 tf_basic3.py，输入以下代码并保存。

```
import tensorflow as tf

a=tf.placeholder(tf.int8,name="input1")
b=tf.placeholder(tf.int8,name="input2")
c=tf.add(a,b,name="my_add_op")
d=tf.multiply(a,c,name="my_mul_op")

store=tf.Variable(0.0,name="var")
inc=store.assign(store+1)
init=tf.global_variables_initializer()

sess=tf.Session()
feed_dict={a:3.0,b:2.0}
out=sess.run(d,feed_dict=feed_dict)
print(out)

sess.run(init)
var_value=sess.run(store)
print(var_value)

sess.run(inc)
var_value=sess.run(store)
print(var_value)

sess.close()
```

（3）运行 tf_basic3.py 文档，运行终端命令：

```
python tf_basic3.py
```

可以看到输出结果为 15,0.0,1.0。

下面为采用上下文管理器机制会话模式的代码。

```
import tensorflow as tf

a=tf.placeholder(tf.int8,name="input1")
b=tf.placeholder(tf.int8,name="input2")
c=tf.add(a,b,name="my_add_op")
```

```
d=tf.multiply(a,c,name="my_mul_op")

store=tf.Variable(0.0,name="var")
inc=store.assign(store+1)
init=tf.global_variables_initializer()

feed_dict={a:3.0,b:2.0}

with tf.Session() as sess:
    sess.run(init)
    out=sess.run(d,feed_dict=feed_dict)
    print(out)
    var_value=sess.run(store)
    print(var_value)
    sess.run(inc)
    var_value=sess.run(store)
    print(var_value)
```

习题 2

2.1 简述 TensorFlow 中 Tensor 和 Flow 代表的含义。
2.2 简述 TensorFlow 计算图中节点和边线所代表的含义。
2.3 编写 TensorFlow 程序分为哪两个阶段？每个阶段各完成什么任务？
2.4 简述基于 virtualenv 安装 TensorFlow 的基本流程。
2.5 在基于 virtualenv 安装方式下如何启动 TensorFlow？
2.6 TensorFlow 运行会话的两种方式是什么？各有什么特点？
2.7 采用上下文会话模式时，如何避免出现 unexpected indent 错误？
2.8 简述 Tensor 的三个属性。
2.9 如何得到 Tensor 的数值？
2.10 TensorFlow 支持的数据类型有哪些？
2.11 试编写一个能够实现任意两个 float32 类型的数据比较大小的 TensorFlow 代码。

第 3 章

TensorFlow 实现神经网络模型训练与测试

本章介绍在 TensorFlow 学习框架下，以 MNIST 手写字体的识别为例，实现 Softmax Regression 简易神经网络及卷积神经网络模型的训练，重点介绍基于 FPGA 实现神经网络中用到的典型 TensorFlow 功能，如数据集格式转换等。

3.1 神经网络训练与测试的基本概念

对于常规的监督机器学习或深度学习，一般分为训练和预测两个阶段。在训练阶段，需要准备原始数据和与之对应的分类标签数据，通过训练得到神经网络模型。在预测阶段，需要使用训练阶段得到的神经网络模型对测试数据进行处理。训练阶段和测试阶段使用的输入数据集合分别称为训练样本集和测试样本集。

在设计人工神经网络模型结构前，需要对输入和输出的数据进行量化。假设输入的数据有 k 个，输出为 n 个分类，那么输入层的神经元节点数应设置为 k，输出层神经元节点数应设置为 n。隐形层的神经元节点数可根据不同的应用需求进行合理设置，隐形层的神经元节点数越多，神经网络的非线性越显著，能够处理的应用越复杂，但会增加计算的复杂度。习惯上，一般第 l 层神经网络的节点数是 $l-1$ 层节点数的 $1\sim1.5$ 倍。

当人工神经网络模型结构定义好后，输入层、隐形层和输出层的节点数就会确定，剩余没有确定的参数还有权值 W 和偏置 b。

3.1.1 神经网络的训练

训练神经网络模型的目的是确定权值 W 和偏置 b，训练的过程也称神经网络的学习过程。神经网络的训练实际上是通过算法不断调整权值 W 和偏置 b，以尽可能地与真实的模型逼近，从而使整个神经网络的预测效果达到最佳。

首先给所有的权值 W 和偏置 b 赋予随机值，使用这些随机生成的权值和偏置参数预测训练样本集合。假设样本的预测值为 y'，真实值为 y。定义一个损失函数衡量预测值和真实值的逼近程度，损失函数越小，神经网络模型的预测效果越好。因此，训练的过程可以理解为调整权值 W 和偏置 b 以使损失函数最小。

损失函数的定义有多种形式,求损失函数最小值是一个优化问题。对神经网络的优化就是使损失函数收敛。为解决该优化问题,目前的方法是使用反向传播算法求得网络模型所有参数的梯度,通过梯度下降算法对网络参数进行更新,当损失函数收敛到一定程度时结束训练,并保存训练结束时的神经网络参数。

3.1.2 神经网络的测试

在训练阶段,通过算法修改神经网络的权值 W 和偏置 b,以使损失函数最小,当损失函数收敛到某个阈值或等于 0 时停止训练,就可以得到训练后的模型的权值和偏置。此时,神经网络的所有参数:输入层、输出层、隐形层、权值和偏置都是已知的。

在测试阶段,将神经网络的测试样本集从输入层开始输入,沿数据流动的方向在网络中进行计算,直到数据从输出层输出,输出层的输出即为预测结果。

3.2 基于 TensforFlow 训练神经网络实现 MNIST 数据集识别

3.2.1 MNIST 数据集

MNIST 数据集来自美国国家标准与技术研究所(National Institute of Standards and Technology,NIST)。训练集(training set)由 250 个不同人手写的数字构成,其中 50% 是高中学生,50% 是人口普查局(the Census Bureau)的工作人员。测试集(test set)也是同样比例的手写数字数据。训练集有 55 000 个样本,测试集有 10 000 个样本,同时验证集有 5000 个样本。每个样本都有与之相对应的标注信息,即 label。MNIST 数据集被分成三部分:55 000 行的训练数据集(mnist.train)、10 000 行的测试数据集(mnist.test)和 5000 行的验证数据集(mnist.validation)。这样的划分很重要,在机器学习模型设计时,必须有一个单独的测试数据集不用于训练,而是用来评估这个模型的性能,从而可以更容易地把设计的模型推广到其他数据集上,使其具有更好的泛化性能。

每个 MNIST 数据样本都由两部分组成:一张包含手写数字的图片和一个对应的标签。训练数据集的图片和标签分别是 mnist.train.images 和 mnist.train.labels。测试数据集的图片和标签分别是 mnist.test.images 和 mnist.test.labels。

所有数据集中的样本均为 28×28 像素的灰度图片,即每张图片中有 28×28=784 个像素点。可以用一个数字数组表示每张图片,如图 3-1 所示。空白部分全部为 0,有笔迹的地方根据颜色深浅在 0~1 之间取值。

把这个 28×28 的矩阵展开成一维数组,数组长度是 28×28=784。如何展开这个数组不重要,只要保持各张图片采用相同的方式展开即可。

注意:展平图片的数字数组会丢失图片的二维结构信息,这显然是不理想的,最优秀的计算机视觉方法会挖掘并利用这些结构信息。此处不需要建立复杂的模型,所以对问题进行了简化,丢弃了空间结构信息。

在 MNIST 训练数据集中,mnist.train.images 是一个形状为[55000,784]的张量

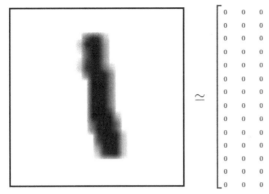 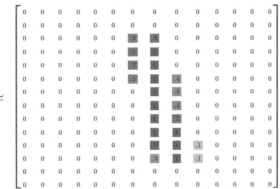

图 3-1　手写数字灰度信息示例

(Tensor)，第一个维度是图片的编号，用来索引图片；第二个维度是图片中像素点的编号，用来索引每张图片中的像素点。因此，mnist.train.images 张量中的每个元素都表示某张图片中的某个像素的强度值，值介于 0 和 1。mnist.train.images 张量中的每张图片都可表示为一个 784 维的向量（如图 3-2 所示）。

相对应的 MNIST 数据集的标签 mnist.train.labels 是一个形状为 [55000, 10] 的张量 (Tensor)，第一个维度是标签的编号，用来索引图片（如图 3-3 所示）。第二个维度是标签的取值，用来表示对应图片代表的数字，是 0～9 的数字。标签数据采用 one-hot 编码，即一个 one-hot 数据除了某一位的数字是 1 以外，其余各维度的数字都是 0。数字 n 将表示成一个只有在第 n 维度（从 0 开始）数字为 1 的 10 维向量。例如，标签 0 将表示成 ([1, 0, 0, 0, 0, 0, 0, 0, 0, 0])。

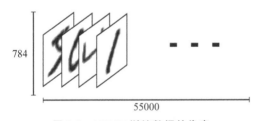 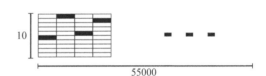

图 3-2　MNIST 训练数据的像素　　　　图 3-3　MNIST 训练数据的标签

3.2.2　Softmax Regression 模型

MNIST 的每张图片都表示一个 0～9 的数字。完成图片识别需要根据给定图片代表的每个数字的概率。例如，模型可能推测一张图片代表数字 9 的概率是 80%，代表数字 8 的概率是 5%，代表其他 8 种数字的概率比 5% 更小，则模型预测该图片代表数字 9。这是一个使用 Softmax Regression 模型的经典案例。Softmax Regression 模型可以用来给不同的对象分配概率。

为了得到一张给定图片属于某个特定数字类的特征，需要对图片像素进行加权求和。如果这个像素具有很强的特征可以说明这张图片不属于该类，那么相应的权值为负数，相反如果这个像素具有很强的特征支持这张图片属于这个类，那么权值是正数。

同时需要加入一个额外的偏置量(bias),因为输入往往会带有一些无关的干扰量。因此对于给定的输入图片 x,它代表数字 i 的判定依据特征可以表示为

$$\text{feature}_i = \sum_j W_{i,j} x_j + b_i$$

x_j 表示图片 x 中的第 j 个像素,$W_{i,j}$ 代表第 j 个像素对应的权重,b_i 代表数字 i 的偏置。$\sum_j W_{i,j} x_j$ 表示对图片中所有像素与对应权值相乘后求和。使用 softmax() 函数可以将这些特征转换成概率:

$$y_i = \text{softmax}(\text{feature}_i)$$

这里的 softmax() 可以看成是一个激活(activation)函数,把定义的线性函数的输出转换成关于 10 个数字类的概率分布。因此,对于一张给定的图片,它与 0~9 的吻合度可以用 softmax() 函数转换成一个概率值。

对于 MNIST 数据集分类,softmax() 函数可以定义为

$$\text{softmax}(x_i) = \text{normlize}(\exp(x_i)) = \frac{\exp(x_i)}{\sum_{k=0}^{9} \exp(x_k)}$$

图 3-4 是 3×3 网络模型结构,可以看出 Softmax Regression 模型是一个没有隐形层的最浅全链接神经网络。根据原理,可以很容易地将该模型扩展为适合 0~9 数字识别的 784×10 网络结构。图 3-5 为图 3-4 网络模型的数学表达,图 3-6 将数学表达以矩阵形式表示。

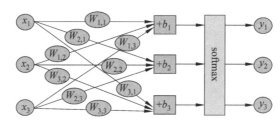

图 3-4 3×3 网络模型结构

$$\begin{bmatrix} y_1 \\ y_2 \\ y_3 \end{bmatrix} = \text{softmax} \begin{pmatrix} W_{1,1} x_1 + W_{1,2} x_1 + W_{1,3} x_1 + b_1 \\ W_{2,1} x_2 + W_{2,2} x_2 + W_{2,3} x_2 + b_2 \\ W_{3,1} x_3 + W_{3,2} x_3 + W_{3,3} x_3 + b_3 \end{pmatrix}$$

图 3-5 3×3 网络模型的数学表达

$$\begin{bmatrix} y_1 \\ y_2 \\ y_3 \end{bmatrix} = \text{softmax} \begin{pmatrix} \begin{bmatrix} W_{1,1} & W_{1,2} & W_{1,3} \\ W_{2,1} & W_{2,2} & W_{2,3} \\ W_{3,1} & W_{3,2} & W_{3,3} \end{bmatrix} \cdot \begin{bmatrix} x_1 \\ x_2 \\ x_3 \end{bmatrix} + \begin{bmatrix} b_1 \\ b_2 \\ b_3 \end{bmatrix} \end{pmatrix}$$

图 3-6 3×3 网络模型的数学表达(矩阵形式)

根据图 3-6，可以分别将输入、输出、权值和偏置用向量的形式表示，则 Softmax Regression 模型的矩阵表达式为

$$y = \text{softmax}(Wx + b)$$

3.2.3 MNIST 数据识别的 Softmax Regression 神经网络模型

1. Softmax Regression 神经网络模型训练与测试的 TensorFlow 实现

按照第 2 章的安装步骤完成 TensorFlow 安装后，在路径 /home/tensorflow/lib/python2.7/site-packages/tensorflow/examples/tutorials/mnist/下有 MNIST 模型的例程。本教材不使用该例程，只使用例程中的个别文件。

（1）MNIST 数据集下载。

下载地址为 http://yann.lecun.com/exdb/mnist/（在网页地址栏中应严格按照下载地址格式输入网址）。

```
train-images-idx3-ubyte.gz:training set images (9912422 bytes)
train-labels-idx1-ubyte.gz: training set labels (28881 bytes)
t10k-images-idx3-ubyte.gz:test set images (1648877 bytes)
t10k-labels-idx1-ubyte.gz: test set labels (4542 bytes)
```

（2）在 Ubuntu 系统下，在/home/tensorflow/文件夹下新建 design 文件夹（若文件夹已存在，则省略该操作），在 design 文件夹下新建文件夹 my_mnist，在 my_mnist 文件夹下新建文件夹 MNIST_data。将下载的包含 4 个压缩包的 MNIST 数据集复制到 MNIST_data 文件夹中。

（3）将/home/tensorflow/lib/python2.7/site-packages/tensorflow/examples/tutorials/mnist/中的 input_data.py 复制到/home/tensorflow/design/my_mnist/目录下。

（4）在/home/tensorflow/design/my_mnist/目录下新建空白文档，并命名为 my_mnist_simple.py，输入以下代码。

```python
#coding: utf-8
import tensorflow as tf
import input_data
import pylab

#载入数据集
mnist=input_data.read_data_sets("MNIST_data",one_hot=True)
#每个批次的大小
batch_size=100
#计算一共有多少个批次
n_batch=mnist.train.num_examples // batch_size
#================定义计算图======================
#定义两个placeholder
x=tf.placeholder(tf.float32,[None,784])
```

```
y=tf.placeholder(tf.float32,[None,10])
#创建一个简单的神经网络
W=tf.Variable(tf.zeros([784,10]))
b=tf.Variable(tf.zeros([10]))
prediction=tf.nn.softmax(tf.matmul(x,W)+b)
#使用交叉熵,梯度下降更有效
loss=tf.reduce_mean(tf.nn.softmax_cross_entropy_with_logits(labels=y,
logits=prediction))
#使用梯度下降法
train_step=tf.train.GradientDescentOptimizer(0.2).minimize(loss)
#初始化变量
init=tf.global_variables_initializer()
#结果存放在一个布尔型列表中
correct_prediction=tf.equal(tf.argmax(y,1),tf.argmax(prediction,1))
#argmax 返回一维张量中最大值所在的位置
#求准确率
accuracy=tf.reduce_mean(tf.cast(correct_prediction,tf.float32))
#=====================定义计算图结束=========================
#=====================运行计算图=========================
with tf.Session() as sess:
    sess.run(init)
    for epoch in range(101):
        for batch in range(n_batch):
            batch_xs,batch_ys= mnist.train.next_batch(batch_size)
            sess.run(train_step,feed_dict={x:batch_xs,y:batch_ys})
            train_acc=sess.run(accuracy,feed_dict=
                    {x:mnist.train.images,y:mnist.train.labels})
        print("Iter "+str(epoch)+",Training Accuracy "+str(train_acc))
        test_acc=sess.run(accuracy,feed_dict=
                    {x:mnist.test.images,y:mnist.test.labels})
    print("Testing Accuracy "+str(test_acc))
#=====================计算图运行结束=========================
```

(5) 安装 pylab 和 python-tk。

```
python -m pip install matplotlib
apt-get install python-tk
```

(6) 运行 my_mnist_simple.py。

```
(tensorflow)$ python my_mnist_simple.py
```

运行结果如图 3-7 所示。

```
Iter 80,Training Accuracy 0.9323636
Iter 81,Training Accuracy 0.93203634
Iter 82,Training Accuracy 0.9322182
Iter 83,Training Accuracy 0.9324
Iter 84,Training Accuracy 0.9326182
Iter 85,Training Accuracy 0.9326
Iter 86,Training Accuracy 0.93274546
Iter 87,Training Accuracy 0.9325455
Iter 88,Training Accuracy 0.93283635
Iter 89,Training Accuracy 0.93314546
Iter 90,Training Accuracy 0.93307275
Iter 91,Training Accuracy 0.9333636
Iter 92,Training Accuracy 0.9332727
Iter 93,Training Accuracy 0.9336
Iter 94,Training Accuracy 0.93341815
Iter 95,Training Accuracy 0.93347275
Iter 96,Training Accuracy 0.9336182
Iter 97,Training Accuracy 0.9337636
Iter 98,Training Accuracy 0.93381816
Iter 99,Training Accuracy 0.93396366
Iter 100,Training Accuracy 0.93389094
Testing Accuracy 0.9292
```

图 3-7　Softmax Regression 神经网络模型的运行结果

2. 代码解读

（1）♯coding：utf-8。

当 py 文件中含有中文字符时，需要在 py 文件的开始执行♯coding：utf-8 命令。

（2）import tensorflow as tf。

在 Python 中采用 import tensorflow as tf 的形式载入 TensofFlow，这样可以使用 tf 代替 tensorflow 作为模块名称，使整个程序更加简洁。

（3）import input_data。

导入 input_data 模块，该模块中定义了读取数据的功能。

（4）import pylab。

导入 pylab 模块。

（5）mnist＝input_data.read_data_sets("MNIST_data",one_hot＝True)。

载入 MNIST 数据集，MNIST_data 为 MNIST 数据集所在的路径，one_hot＝True 表示使用 one_hot 编码格式。

（6）batch_size＝100。

定义每批次的训练样本数为 100。

（7）n_batch＝mnist.train.num_examples// batch_size。

根据训练样本集和每批次样本数量计算批次数量。mnist.train.num_examples＝55000，"//"代表除法运算符号。

（8）x＝tf.placeholder(tf.float32,[None,784])。

为样本图片数据创建一个 placeholder，即输入数据的地方，第一个参数代表数据类型，第二个参数代表 Tensor 的 shape，即数据的尺寸，None 代表不限输入样本的数量，784 表示每个样本是一个 784 维的向量。

（9）y＝tf.placeholder(tf.float32,[None,10])。

为样本标签数据创建一个 placeholder。

(10) W=tf.Variable(tf.zeros([784,10]))。

为神经网络的权值创建一个 Variable,尺寸为 784×10,784 对应输入图片的特征维数,10 对应 10 类数字。Variable 在模型训练迭代中一直存在并在每轮迭代中进行更新。

(11) b=tf.Variable(tf.zeros([10]))。

为神经网络的偏置创建一个 Variable,尺寸为 10,10 对应 10 类数字。

(12) prediction=tf.nn.softmax(tf.matmul(x,W)+b)。

tf.nn.softmax 定义了 Softmax Regression 模型,softmax 是 tf.nn 下面的一个函数。tf.nn 包含大量神经网络的组件,包括卷积操作 conv、池化操作 pooling、归一化、loss、分类操作等。tf.matmul 是 TensorFlow 中的矩阵乘法函数。

(13) loss=tf.reduce_mean(tf.nn.softmax_cross_entropy_with_logits(labels=y,logits=prediction))。

为了训练模型,定义一个 loss function(损失函数)描述模型对问题的分类精度,loss 越小,代表模型的分类效果与真实值的偏差越小,即模型越准确。训练的目的是不断将这个 loss 减小,直至达到一个全局最优解或局部最优解。

对于多分类问题,通常使用 cross-entropy 作为 loss function,定义表达式如下:$H_{y'}(y) = -\sum_i y'_i \log(y_i)$,其中 y 是预测的概率分布,y′是真实的概率分布(即 label 的 one-hot 编码),通常可以用该 loss function 判断模型对真实概率分布估计的准确程度。注意:在本例程中,y 代表真实的概率分布,prediction 代表预测的概率分布,因此 loss function 的表达式应修改为 $H_y(prediction) = -\sum_i y_i \log(prediction_i)$。

tf.nn.softmax_cross_entropy_with_logits 函数所做的操作是将每个样本真实标签的 one-hot 向量与其预测概率的对数相乘,将结果按元素取负号,然后按样本将元素逐行相加。

tf.reduce_mean 函数是求均值。

只要定义 loss,训练时就会自动求导并进行梯度下降,完成对 Softmax Regression 模型参数的自动学习。

(14) train_step=tf.train.GradientDescentOptimizer(0.2).minimize(loss)。

采用随机梯度下降 SGD(Stochastic Gradient Descent)优化算法,定义优化算法后,TensorFlow 就可以根据定义的整个计算图(代码中定义的公式可以自动构成计算图)自动求导,并根据反向传播算法进行训练,在每一轮迭代时更新参数以减小 loss。

tf.train.GradientDescentOptimizer 是 TensorFlow 提供的一个封装好的优化器,自动添加许多运算操实现反向传播和梯度下降,用户只需要在每轮迭代时 feed 数据即可。在本例程中,0.2 为学习速率,优化目标设置为 loss,train_step 为定义的训练操作。

(15) init=tf.global_variables_initializer()。

定义初始化操作 init,运行该操作可以对程序中的所有参数进行初始化。tf.global_variables_initializer()为 TensorFlow 的全局参数初始化函数。

(16) correct_prediction=tf.equal(tf.argmax(y,1),tf.argmax(prediction,1))。

tf.argmax 是从一个 tensor 中寻找最大值的序号,tf.argmax(y,1)得到样本真实数字的类别,tf.argmax(prediction,1)得到各个预测的数字中概率最大的那一个,tf.equal 判断 tf.argmax(y,1)和 tf.argmax(prediction,1)是否相等,当二者相等时,correct_prediction 为真,否则 correct_prediction 为假。

(17) accuracy=tf.reduce_mean(tf.cast(correct_prediction,tf.float32))。

定义全部样本预测的准确率。tf.cast 实现类型转换,将 correct_prediction 由 bool 类型转换为 float32 类型。

至此,我们使用 TensorFlow 实现了一个简单的机器学习算法模型 Softmax Regression,(1)~(17)完成了计算图的定义。接下来就是运行计算图,进行迭代训练和测试。

(18) with tf.Session() as sess。

创建一个名为 sess 的会话,并通过 Python 中的上下文管理这个会话。通过 Python 上下文管理机制,只要将所有的计算放在 with 内部即可,当上下文管理器退出时,会自动释放所有的资源,这样既解决了异常退出时资源释放的问题,又解决了忘记调用 Session.close 函数而产生的资源泄露。

(19) sess.run(init)。

使用会话执行 init 操作,实现所有参数的初始化。

(20) for epoch in range(101)。

for 循环控制语句,epoch 为循环变量,取值范围为 0~100。

(21) for batch in range(n_batch)。

for 循环控制语句,batch 为循环变量,取值范围为 0~n_batch-1。

(22) batch_xs,batch_ys=mnist.train.next_batch(batch_size)。

每批次从训练集中一次性提取 batch_size 张图片和对应的标签。将图片赋值给 batch_xs,将标签赋值给 batch_ys。

(23) sess.run(train_step,feed_dict={x:batch_xs,y:batch_ys})。

使用会话执行 train_step 操作,即 train_step 是运行结果的输出张量。输入数据 x 为 batch_xs,y 为 batch_ys。该代码的作用是实现神经网络训练,不断调整网络参数,以使 loss 最小。

(24) train_acc = sess.run(accuracy,feed_dict = {x:mnist.train.images,y:mnist.train.labels})。

使用会话执行 accuracy 操作,即 accuracy 是运行结果的输出张量。输入数据 x 为 mnist.train.images,y 为 mnist.train.labels。该代码的作用是用训练样本集验证当前网络参数的识别准确率。

(25) print("Iter"+str(epoch)+",Training Accuracy"+str(train_acc))。

打印每次 epoch 的训练样本的识别结果信息,需要将打印的参数用 str 转换为 string 类型。

(26) test_acc=sess.run(accuracy,feed_dict={x:mnist.test.images,y:mnist.test.labels})。

使用会话执行 accuracy 操作,即 accuracy 是运行结果的输出张量。输入数据 x 为

mnist.test.images,y 为 mnist.test.labels。该代码的作用是用测试样本集验证当前网络参数的识别准确率。

(27) print("Testing Accuracy"＋str(test_acc))。

打印测试样本的识别结果信息,需要将打印的参数用 str 转换为 string 类型。

注意：若提示 unexpected indent 错误信息,则说明代码中行之间的缩进不一致。

3.2.4　MNIST 数据识别的卷积神经网络模型

卷积神经网络(convolutional neural networks,CNN)已成为众多科学领域的研究热点,特别是在模式分类领域,由于卷积神经网络避免了对图像复杂的前期预处理,可以直接输入原始图像,因此得到了更为广泛的应用。

1. 卷积神经网络简介

局部感知、权值共享和池化是卷积神经网络的三个重要特征。

1) 局部感知

当图像的尺寸很大时,全连接神经网络的参数将变得很多,从而导致计算量非常大。为了降低计算量,卷积神经网络采用局部感知的方法,即每个神经元只感受局部的图像区域,将这些不同局部的神经元综合起来就可以得到全局信息,如图 3-8 所示。

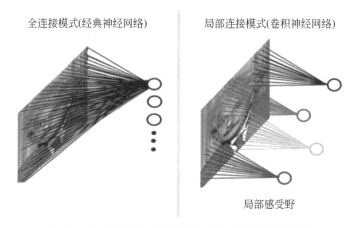

图 3-8　全连接网络全面感知与 CNN 局部感知的对比

感受野(receptive field)是 CNN 每一层输出的特征图(feature map)上每个像素点(神经元)在该层输入图像上映射的区域大小。感受野也可以认为是特征图中的神经元"看到的"输入区域,特征图上某个神经元的计算只受该神经元所看到的输入区域的影响,与区域之外的内容没有关系。图 3-9 展示了卷积神经网络中的输入图像、卷积核、感受野、特征图等基本术语,以及输入图像与卷积和进行卷积操作得到特征图的基本原理。图 3-9 中感受野的尺寸为 3×3,与卷积核尺寸相同,感受野与卷积核对应位置的元素相乘后累加,将得到新的元素构成特征图。

特征图的大小(卷积特征)由 3 个参数控制：深度(depth)、步长(stride)、填充

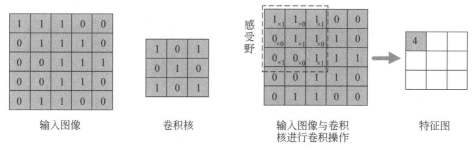

图 3-9　CNN 中的基本术语图示

（padding）。深度是指卷积操作所需的滤波器个数。步长是指每次卷积运算结束后至下一次卷积运算开始前，卷积核在输入图像上滑动的像素数。当步长为 $n(n\geqslant 1$ 的整数）时，卷积核每滑动一次，新的感受野与前一个感受野相比位置向右平移 n 个像素单位，若本行结束，则向下平移 n 个像素单位，从最左侧重复滑动操作。填充可以解决图像边界卷积造成图像信息丢失的问题，主要有 3 种方法：Same Padding、Valid Padding 和自定义 Padding。Same Padding 根据卷积核的大小对输入图像进行边界补充，一般填充 0 值，以使卷积后得到的特征图与输入图像的尺寸相同；Valid Padding 保持输入图像尺寸不变，即不进行填充；自定义 Padding 与生成特征图尺寸的计算表达式为

$$\text{output}_h = (\text{input}_h + 2\times \text{padding}_h - \text{kernel}_h)/\text{stride} + 1 \quad (3\text{-}1)$$

$$\text{output}_w = (\text{input}_w + 2\times \text{padding}_w - \text{kernel}_w)/\text{stride} + 1 \quad (3\text{-}2)$$

式中，input、output 分别为输入图像和特征图的尺寸，kernel 为卷积核的尺寸，stride 为步长，padding 为边界填充尺寸，h、w 分别为输入图像的长和宽。

图 3-10 为 Valid Padding，stride=1、Valid Padding，stride=2 以及 Same Padding，stride=1 的示例。读者可以根据式（3-1）和式（3-2），自行推导 Same Padding，stride=2 时的情况。

2）权值共享

卷积核的局部感知机制将输入图像划分为不同区域（不同的感受野），但不同的感受野与具有相同数值的卷积核进行特征图计算，即图中每个感受野会被同样的卷积核处理。卷积核中的数值称作权值（权重），因此称为权值共享。如图 3-11 所示，输入图像的不同区域用同一个卷积核进行卷积运算可以得到特征图输出。

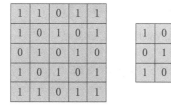

(a) 输入图像和卷积核

图 3-10　padding 模式和步长应用示例

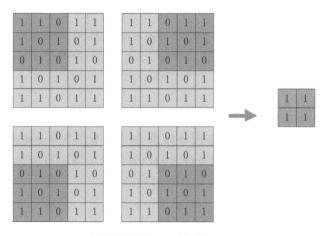

(b) Valid Padding, stride=1

(c) Valid Padding, stride=2

图 3-10 （续）

(d) Same Padding, stride=1

图 3-10 （续）

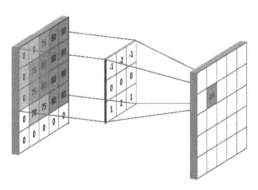

图 3-11　CNN 中的权值共享

3）池化

池化也称下采样，是指将特征图中的若干元素根据池化规则合并为一个元素的过程。池化可以看作是池化规则对特征图的滑动滤波，如图 3-12 所示。基本池化规则主要有两种：取最大值和求平均值。因此池化方式也有两种：最大池化和均值池化。图3-13为最大池化和平均池化的例子。

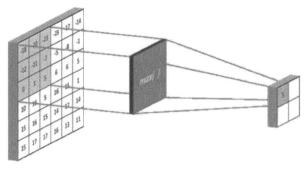

图3-12 池化操作原理

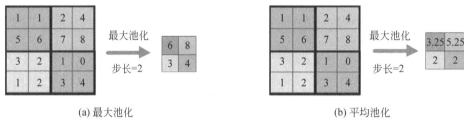

(a) 最大池化　　　　　　　　　　　(b) 平均池化

图3-13 池化示例

2. 卷积神经网络的 TensorFlow 实现

1) 卷积神经网络结构

两个卷积层+池化层，最后接上两个全连接层。第一层卷积使用32个3×3×1的卷积核，步长为1，边界处理方式为same padding，即卷积的输入和输出保持相同的尺寸。激活函数为ReLU（Rectified Linear Unit），后接一个2×2的池化层，方式为最大池化。ReLU的数学表达如式(3-3)所示，图3-14给出了ReLU函数的曲线表示。

$$f(x) = \max(x, 0) \tag{3-3}$$

第二层卷积使用50个3×3×32的卷积核，步长为1，边界处理方式为same padding，激活函数为ReLU，后接一个2×2的池化层，方式为最大池化。

第一层全连接层使用1024个神经元，激活函数依然是ReLU。

第二层全连接层使用10个神经元，激活函数为softmax，用于输出结果。

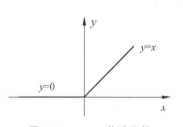

图3-14 ReLU激活函数

2) TensorFlow 代码

(1) 在/home/tensorflow/design/my_mnist/目录下新建空白文档，命名为 my_mnist_cnn.py，输入以下代码：

```python
#coding=utf-8
import input_data
import tensorflow as tf
import pylab
#读取数据
mnist=input_data.read_data_sets('MNIST_data', one_hot=True)
#每个批次的大小
batch_size=100
#计算一共有多少个批次
n_batch=mnist.train.num_examples // batch_size
#==============构建CNN网络结构开始=============
#自定义卷积函数
def conv2d(x,w):
    return tf.nn.conv2d(x,w,strides=[1,1,1,1],padding='SAME')
#自定义池化函数
def max_pool_2_2(x):
    return tf.nn.max_pool(x,ksize=[1,2,2,1],strides=[1,2,2,1],padding='SAME')
#设置占位符,尺寸为样本输入和输出的尺寸
x=tf.placeholder(tf.float32,[None,784])
y_=tf.placeholder(tf.float32,[None,10])
x_img=tf.reshape(x,[-1,28,28,1])
#设置第一个卷积层和池化层
w_conv1=tf.Variable(tf.truncated_normal([3,3,1,32],stddev=0.1))
#32个3×3×1的卷积核
b_conv1=tf.Variable(tf.constant(0.1,shape=[32])) #32个通道
h_conv1=tf.nn.relu(conv2d(x_img,w_conv1)+b_conv1)
h_pool1=max_pool_2_2(h_conv1)
#设置第二个卷积层和池化层
w_conv2=tf.Variable(tf.truncated_normal([3,3,32,50],stddev=0.1))
#50个3×3×32的卷积核
b_conv2=tf.Variable(tf.constant(0.1,shape=[50])) #50个通道
h_conv2=tf.nn.relu(conv2d(h_pool1,w_conv2)+b_conv2)
h_pool2=max_pool_2_2(h_conv2)
#设置第一个全连接层
w_fc1=tf.Variable(tf.truncated_normal([7*7*50,1024],stddev=0.1))
b_fc1=tf.Variable(tf.constant(0.1,shape=[1024]))
h_pool2_flat=tf.reshape(h_pool2,[-1,7*7*50])
h_fc1=tf.nn.relu(tf.matmul(h_pool2_flat,w_fc1)+b_fc1)
h_fc1_drop=tf.nn.dropout(h_fc1,0.5)
#设置第二个全连接层
w_fc2=tf.Variable(tf.truncated_normal([1024,10],stddev=0.1))
b_fc2=tf.Variable(tf.constant(0.1,shape=[10]))
y_out=tf.nn.softmax(tf.matmul(h_fc1_drop,w_fc2)+b_fc2)
```

```
#==============构建 CNN 网络结构结束==============
#使用交叉熵
loss=tf.reduce_mean(tf.nn.softmax_cross_entropy_with_logits(labels=y_,
logits=y_out))
#使用梯度下降法
train_step=tf.train.GradientDescentOptimizer(0.2).minimize(loss)
#定义初始化操作
init=tf.global_variables_initializer()
#建立正确率计算表达式
correct_prediction=tf.equal(tf.argmax(y_out,1),tf.argmax(y_,1))
accuracy=tf.reduce_mean(tf.cast(correct_prediction,tf.float32))
#开始训练
with tf.Session() as sess:
    sess.run(init)
    for epoch in range(101):
        for batch in range(n_batch):
            batch_xs,batch_ys=mnist.train.next_batch(batch_size)
            sess.run(train_step,feed_dict={x:batch_xs,y_:batch_ys})
            train_acc=sess.run(accuracy,feed_dict={x:mnist.train.images,
                    y_:mnist.train.labels})
        print("Iter "+str(epoch)+",Training Accuracy "+str(train_acc))
    test_acc=sess.run(accuracy,feed_dict={x:mnist.test.images,y_:mnist.
    test.labels})
    print("Testing Accuracy "+str(test_acc))
```

（2）运行 my_mnist_cnn.py，在目录/home/tensorflow/design/my_mnist/下运行 my_mnist_cnn.py。

```
(tensorflow)$ python my_mnist_cnn.py
```

图 3-15 为卷积神经网络模型的运行结果，显示测试结果为 test_accuracy=0.9907。

```
Iter 80,Training Accuracy 0.9976
Iter 81,Training Accuracy 0.99774545
Iter 82,Training Accuracy 0.99756366
Iter 83,Training Accuracy 0.99783635
Iter 84,Training Accuracy 0.99767274
Iter 85,Training Accuracy 0.99783635
Iter 86,Training Accuracy 0.9976909
Iter 87,Training Accuracy 0.9979454
Iter 88,Training Accuracy 0.99785453
Iter 89,Training Accuracy 0.9976364
Iter 90,Training Accuracy 0.9975455
Iter 91,Training Accuracy 0.9980364
Iter 92,Training Accuracy 0.9980909
Iter 93,Training Accuracy 0.99807274
Iter 94,Training Accuracy 0.9978909
Iter 95,Training Accuracy 0.99774545
Iter 96,Training Accuracy 0.9980182
Iter 97,Training Accuracy 0.9982182
Iter 98,Training Accuracy 0.9978182
Iter 99,Training Accuracy 0.9982182
Iter 100,Training Accuracy 0.9981818
Testing Accuracy 0.9907
```

图 3-15　卷积神经网络模型的运行结果

对比图 3-15 和图 3-7 可以发现,与 Softmax Regression 神经网络模型进行比较,采用结构更复杂的卷积神经网络可以实现更高的识别正确率。

3. 代码解读

本部分将对 my_mnist_cnn.py 代码中出现的新的 API 函数进行介绍。未解读的代码请参考 3.2.3 节中的代码解读中对 my_mnist_simple.py 的分析。

(1) tf.nn.conv2d 是 TensorFlow 专门用作卷积运算的 API 函数,该函数的原型为:

```
tf.nn.conv2d(input, filter, strides, padding, use_cudnn_on_gpu=None, name=None)
```

该 API 函数各参数的意义如下。

① input:输入图像或数据,是一个 Tensor,shape=(4,),shape 格式为 1×4,4 个元素为[batch, in_height, in_width, in_channels],具体含义是[一个 batch 的图片数量,图片高度,图片宽度,图像通道数],dtype 为 float32 或 float64。

② filter:CNN 的卷积核,也是一个 Tensor,shape=(4,),shape 格式为 1×4,4 个元素为[filter_height, filter_width, in_channels, out_channels],具体含义是[卷积核高度,卷积核宽度,图像通道数,卷积核个数],dtype 与参数 input 相同。第 3 个元素 in_channels 与 input 的第 4 个元素 in_channels 相同。

③ strides:卷积运算时在图像每一维的步长,是一个一维向量,长度为 4,即含有 4 个元素,格式为[1, 横向步长, 纵向步长, 1]。

④ padding:只能是 SAME 和 VALID 其中之一。

⑤ use_cudnn_on_gpu:bool 类型,指是否使用 cudnn 加速,默认为 true。

⑥ name:指定本次调用 API 函数操作的名字。

该 API 会返回一个 Tensor,即卷积神经网络中的 feature map,shape 格式为[batch, height, width, channels]。

(2) tf.nn.max_pool 是 TensorFlow 专门用作池化运算的 API 函数,该函数的原型为:

```
tf.nn.max_pool(value, ksize, strides, padding, name=None)
```

该 API 函数各参数的意义如下。

① value:池化输入,一般池化层接在卷积层后面,所以输入通常是 feature map,shape 格式依然是[batch, height, width, channels]。

② ksize:池化窗口尺寸,是一个含有 4 个元素的一维向量,格式为[1, height, width, 1]。首尾两个元素为 1,代表不在 batch 和 channels 上做池化。本例中 ksize 取值为[1, 2, 2, 1],代表横向两个和纵向两个共 4 个特征图中数据为一组进行池化。

③ strides:池化窗口在特征图的每一个维度上滑动的步长,格式为[1, stride, stride, 1]。

④ padding:和卷积类似,可以取 VALID 或者 SAME。

⑤ name:指定本次调用 API 函数操作的名字。

该 API 返回一个 Tensor,类型与输入相同,shape 格式为[batch, height, width,

channels]。

（3）tf.reshape 是 TensorFlow 中用来改变张量 shape 的 API 函数，该函数的原型为：

```
tf.reshape(tensor,shape,name=None)
```

该 API 函数各参数的意义如下。

tensor：输入张量。

shape：变形后得到的输出张量的形状。

name：指定本次调用 API 函数操作的名字。

该 API 函数的作用是将输入 tensor 变换为参数 shape 形式，参数 shape 为一个列表形式，如本例中的[−1,28,28,1]。−1 代表不用指定这一维的大小，函数会自动进行计算，但是列表中只能存在一个−1。如果 batch 为 1，则每次输入一张图像，输入 x 的 shape 为[1,784]，则 x_img 的 shape 为[1,28,28,1]。如果 batch 为 100，则每次输入 100 张图像，输入 x 的 shape 为[100,784]，则 x_img 的 shape 为[100,28,28,1]。

（4）tf.truncated_normal 是 TensorFlow 中用来生成截断正态分布随机数据的 API 函数，该函数的原型为：

```
tf.truncated_normal(shape, mean=0.0, stddev=1.0,dtype=tf.float32,seed=None,name=None)
```

该 API 函数各参数的意义如下。

① shape：表示生成张量的维度。

② mean：均值，默认值为 0.0。

③ stddev：标准差，默认值为 1.0。

④ dtype：数据类型，默认为 tf.float32。

⑤ seed：随机数种子。

⑥ name：指定本次调用 API 函数操作的名字。

该 API 函数根据设定的均值和标准差产生正态分布的随机数据，如果产生正态分布的数值与均值的差值大于标准差的 2 倍，则重新生成，因此称为截断的正态分布生成函数，该函数可保证产生的随机数与均值的差距不会超过标准差的 2 倍。

（5）tf.nn.relu 实现 ReLU 激活函数功能。

（6）tf.nn.dropout 是 TensorFlow 中用来随机屏蔽某些神经元输出的 API 函数，该函数的原型为：

```
tf.nn.dropout(x, keep_prob, noise_shape=None,seed=None,name=None)
```

该 API 函数的各参数意义如下。

① x：输入，输入 tensor。

② keep_prob：float 类型，每个元素被保留下来的概率，设置神经元被选中的概率。

③ noise_shape：一个一维的 int32 张量，代表随机产生"保留/丢弃"标志的 shape。

④ seed：整型变量，随机数种子。

⑤ name：指定本次调用 API 函数操作的名字。

该 API 函数可以在不同的训练过程中按设置的概率随机屏蔽一部分神经元的输出，即让某些神经元的激活值以一定的概率 1-keep_prob 让其停止工作。

各个列表中的取值，与输入图像尺寸、卷积核参数、池化参数等有关，读者可自行推算列表参数中的取值依据。

3.3 MNIST 数据集转换

之所以进行数据集转换，是因为神经网络需要处理的样本集可能是多种多样的，有的是一维时间序列数据，有的是二维图像数据。为了能够实现不同格式样本集的处理，需要将样本集转换为神经网络可以识别的数据格式。通过学习数据集转换，读者可以转换自己采集的样本集的数据格式，然后用相同的方法采用神经网络进行训练和测试。

3.3.1 将数据集转换为以 txt 文件保存的数据

本例将 101 张 MNIST 测试样本和对应的标签转换为 txt 文档。

（1）在/home/tensorflow/design/my_mnist/目录下新建空白文档，命名为 txt_file_gen.py，输入以下代码。

```
#coding: utf-8
import tensorflow as tf
import input_data
import numpy as np
#载入数据集
mnist=input_data.read_data_sets("MNIST_data",one_hot=True)
with tf.Session() as sess:
    for i in range(101):
        img_in_i=mnist.test.images[i]
        tag_i=np.argmax(mnist.test.labels[i])
    np.savetxt('/home/tensorflow/design/my_mnist/mnist_txt/mnist_img_txt/img_%d.txt'%i, img_in_i.reshape(-1), fmt="%.31f",delimiter=",")

    np.savetxt('/home/tensorflow/design/my_mnist/mnist_txt/mnist_lab_txt/img_lab_%d.txt'%i, tag_i.reshape(-1), fmt="%.31f",delimiter=",")
print("text write was sucessful")
```

（2）在/home/tensorflow/design/my_mnist/目录下新建文件夹 mnist_txt，并在该文件夹下新建两个文件夹 mnist_img_txt 和 mnist_lab_txt。

注意：必须提前建立文件夹，Python 代码不能自动生成文件夹。

（3）运行 txt_file_gen.py。

```
(tensorflow)$ python txt_file_gen.py
```

若成功运行,则/home/tensorflow/design/my_mnist/mnist_txt/mnist_img_txt 和 /home/tensorflow/design/my_mnist/mnist_txt/mnist_lab_txt 分别会有 101 个 txt 文件生成。

3.3.2 将数据集转换为以 bmp 文件保存的图片

1. 训练样本集转换

(1) 在/home/tensorflow/design/目录下新建文件夹 my_mnist_tfrecord。

(2) 将/home/tensorflow/design/my_mnist 目录下的 MNIST_data 和 input_data.py 复制到/home/tensorflow/design/my_mnist_tfrecord 文件夹。

(3) 在/home/tensorflow/design/my_mnist_tfrecord/目录下新建空白文档,命名为 mnist_bmp_train_gen.py,输入以下代码。

```
#coding: utf-8
import os
import tensorflow as tf
import input_data
from PIL import Image
#声明图片的宽和高
rows=28
cols=28
#当前路径下的保存目录
save_dir="./mnist_bmp_train"
#读入 MNIST 数据
mnist=input_data.read_data_sets("MNIST_data/", one_hot=False)
#创建会话
sess=tf.Session()
#获取图片总数
shape=sess.run(tf.shape(mnist.train.images))
images_count=shape[0]
pixels_per_image=shape[1]
#获取标签总数
shape=sess.run(tf.shape(mnist.train.labels))
labels_count=shape[0]
#要提取的图片数量
images_to_extract=shape[0]
#
labels=mnist.train.labels
#检查数据集是否符合预期格式
if (images_count==labels_count) and (shape.size==1):
    print ("数据集总共包含 %s 张图片,和 %s 个标签" %(images_count, labels_count))
    print ("每张图片包含 %s 个像素" %(pixels_per_image))
    print ("数据类型: %s" %(mnist.train.images.dtype))
```

```python
#mnist 图像数据的数值范围是[0,1],需要扩展到[0,255],以便于人眼观看
if mnist.train.images.dtype=="float32":
    print ("准备将数据类型从[0,1]转换为binary[0,255]...")
    for i in range(0,images_to_extract):
        for n in range(pixels_per_image):
            if mnist.train.images[i][n]!=0:
                mnist.train.images[i][n]=255
        #打印转换进度
        if ((i+1)%50)==0:
            print ("图像浮点数值扩展进度:已转换 %s 张,共需转换 %s 张" %(i+1,
            images_to_extract))

    #创建数字图片的保存目录
    for i in range(10):
        dir="%s/%s/" %(save_dir,i)
        if not os.path.exists(dir):
            print ("目录 ""%s"" 不存在! 自动创建该目录..." %dir)
            os.makedirs(dir)

    #通过 Python 图片处理库生成图片
    indices=[0 for x in range(0, 10)]
    for i in range(0,images_to_extract):
        img=Image.new("L",(cols,rows))
        for m in range(rows):
            for n in range(cols):
                img.putpixel((n,m), int(mnist.train.images[i][n+m*cols]))
        #根据图片代表的数字label生成对应的保存路径
        digit=labels[i]
        path="%s/%s/%s.bmp" %(save_dir, labels[i], indices[digit])
        indices[digit]+=1
        img.save(path)
        #打印保存进度
        if ((i+1)%50)==0:
            print ("图片保存进度:已保存 %s 张,共需保存 %s 张" %(i+1, images_to_
            extract))

else:
    print ("图片数量和标签数量不一致!")
```

(4) 安装 PIL module,运行以下终端命令。

```
pip install Pillow
```

(5) 运行 mnist_bmp_train_gen.py,运行以下终端命令。

```
python mnist_bmp_train_gen.py
```

运行完毕后,在/home/tensorflow/design/my_mnist_tfrecord/目录下生成一个名为mnist_bmp_train 的文件夹,打开 mnist_bmp_train 文件夹可以看到 10 个子文件夹,名称为 0～9。在每个子文件夹下均有若干后缀为 bmp 的图片。

2. 测试样本集转换

(1) 在/home/tensorflow/design/my_mnist_tfrecord/目录下新建空白文档,命名为 mnist_bmp_test_gen.py,输入以下代码。

```
#coding: utf-8
import os
import tensorflow as tf
import input_data
from PIL import Image
#声明图片的宽和高
rows=28
cols=28
#当前路径下的保存目录
save_dir="./mnist_bmp_test"
#读入 MNIST 数据
mnist=input_data.read_data_sets("MNIST_data/", one_hot=False)
#创建会话
sess=tf.Session()
#获取图片总数
shape=sess.run(tf.shape(mnist.test.images))
images_count=shape[0]
pixels_per_image=shape[1]
#获取标签总数
shape=sess.run(tf.shape(mnist.test.labels))
labels_count=shape[0]
#要提取的图片数量
images_to_extract=shape[0]
#
labels=mnist.test.labels

#检查数据集是否符合预期格式
if (images_count==labels_count) and (shape.size==1):
    print ("数据集总共包含 %s 张图片,和 %s 个标签" %(images_count, labels_count))
    print ("每张图片包含 %s 个像素" %(pixels_per_image))
    print ("数据类型: %s" %(mnist.test.images.dtype))

    #MNIST 图像数据的数值范围是[0,1],需要扩展到[0,255],以便于人眼观看
```

```
        if mnist.test.images.dtype=="float32":
            print ("准备将数据类型从[0,1]转为 binary[0,255]...")
            for i in range(0,images_to_extract):
                for n in range(pixels_per_image):
                    if mnist.test.images[i][n]!=0:
                        mnist.test.images[i][n]=255
                #打印转换进度
                if ((i+1)%50)==0:
                    print ("图像浮点数值扩展进度：已转换 %s 张,共需转换 %s 张" %(i+1,
                        images_to_extract))

        #创建数字图片的保存目录
        for i in range(10):
            dir="%s/%s/" %(save_dir,i)
            if not os.path.exists(dir):
                print ("目录 ""%s"" 不存在！自动创建该目录..." %dir)
                os.makedirs(dir)

        #通过 Python 图片处理库生成图片
        indices=[0 for x in range(0, 10)]
        for i in range(0,images_to_extract):
            img=Image.new("L",(cols,rows))
            for m in range(rows):
                for n in range(cols):
                    img.putpixel((n,m), int(mnist.test.images[i][n+m*cols]))
            #根据图片代表的数字 label 生成对应的保存路径
            digit=labels[i]
            path="%s/%s/%s.bmp" %(save_dir, labels[i], indices[digit])
            indices[digit]+=1
            img.save(path)
            #打印保存进度
            if ((i+1)%50)==0:
                print ("图片保存进度：已保存 %s 张,共需保存 %s 张" %(i+1, images_to_
                    extract))

    else:
        print ("图片数量和标签数量不一致！")
```

（2）运行 mnist_bmp_test_gen.py。

```
python mnist_bmp_test_gen.py
```

运行完毕后，在/home/tensorflow/design/my_mnist_tfrecord/目录下生成一个名为 mnist_bmp_test 的文件夹，打开 mnist_bmp_test 文件夹可以看到 10 个子文件夹，名称为

0~9。在每个子文件夹下均有若干后缀为 bmp 的图片。

3.3.3 将 bmp 转换为 tfrecords 格式

1. 训练样本集转换

将 3.3.2 节得到的 bmp 图片转换为 tfrecords 格式。

（1）生成 train.tfrecord。

修改文件夹 mnist_bmp_train 的访问属性，使该文件夹可写，以方便创建文档，运行以下终端命令。

```
chmod -R 777 /home/tensorflow/design/my_mnist_tfrecord/mnist_bmp_train/
```

在/home/tensorflow/design/my_mnist_tfrecord/mnist_bmp_train/目录下新建空白文档，并命名为 train_tfrecord_gen.py，输入以下代码。

```
#coding=utf-8
import os
import tensorflow as tf
from PIL import Image
import matplotlib.pyplot as plt
import numpy as np

import pylab
cwd='./'
classes={'0','1','2','3','4','5','6','7','8','9'}
writer=tf.python_io.TFRecordWriter("train.tfrecords")
for index, name in enumerate(classes):
    class_path=cwd+name+'/'
    for img_name in os.listdir(class_path):
        img_path=class_path+img_name
    img=Image.open(img_path)
    img=img.resize((28,28))
    img_raw=img.tobytes()
        example=tf.train.Example(features=tf.train.Features(feature={
        "label":tf.train.Feature(int64_list=tf.train.Int64List(value=
        [index])),'img_raw':tf.train.Feature(bytes_list=tf.train.BytesList
        (value=[img_raw]))
        }))
        writer.write(example.SerializeToString())

writer.close()
```

（2）运行 train_tfrecord_gen.py。

```
python train_tfrecord_gen.py
```

运行结束后,在/home/tensorflow/design/my_mnist_tfrecord/mnist_bmp_train/目录下生成名为 train.tfrecords 的文件。

2. 测试样本集转换

(1) 生成 test.tfrecord。

修改文件夹 mnist_bmp_test 的访问属性,使该文件夹可写,以方便创建文档。运行以下终端命令。

```
chmod -R 777 /home/tensorflow/design/my_mnist_tfrecord/mnist_bmp_test/
```

在/home/tensorflow/design/my_mnist_tfrecord/mnist_bmp_test/目录下新建空白文档,命名为 test_tfrecord_gen.py,输入以下代码。

```
#coding=utf-8
import os
import tensorflow as tf
from PIL import Image
import matplotlib.pyplot as plt
import numpy as np

import pylab
cwd='./'
classes={'0','1','2','3','4','5','6','7','8','9'}
writer=tf.python_io.TFRecordWriter("test.tfrecords")

for index, name in enumerate(classes):
    class_path=cwd+name+'/'
    for img_name in os.listdir(class_path):
        img_path=class_path+img_name
    img=Image.open(img_path)
    img=img.resize((28,28))
    img_raw=img.tobytes()
        example=tf.train.Example(features=tf.train.Features(feature={
        "label":tf.train.Feature(int64_list=tf.train.Int64List(value=
        [index])), 'img_raw':tf.train.Feature(bytes_list=tf.train.BytesList
        (value=[img_raw]))
        }))
        writer.write(example.SerializeToString())

writer.close()
```

(2) 运行 test_tfrecord_gen.py。

```
python test_tfrecord_gen.py
```

运行结束后，在/home/tensorflow/design/my_mnist_tfrecord/mnist_bmp_test/目录下生成名为 test.tfrecords 的文件。

（3）在/home/tensorflow/design/my_mnist_tfrecord/目录下新建空白文档并命名为 read_and_decode.py，并输入以下代码。

```
import tensorflow as tf
def read_and_decode(filename):#read .tfrecords
    filename_queue=tf.train.string_input_producer([filename]) # create
    a queue

    reader=tf.TFRecordReader()
    _, serialized_example=reader.read(filename_queue) # return filename
    and file
    features=tf.parse_single_example(serialized_example,
                    features={
                        'label':tf.FixedLenFeature([], tf.int64),
                        'img_raw':tf.FixedLenFeature([], tf.string),
                    })#take out image and label
    img=tf.decode_raw(features['img_raw'],tf.uint8)
    img=tf.reshape(img, [28, 28]) #reshape to
    img=tf.cast(img, tf.float32) * (1./255) - 0.5 #throw out the img tensor
    label=tf.cast(features['label'], tf.int32)
    return img, label
```

3.4 读取 tfrecords 格式数据实现 MNIST 手写字体识别

3.4.1 Softmax Regression 模型

（1）在/home/tensorflow/design/my_mnist_tfrecord/目录下新建空白文档并命名为 my_mnist_tfrecord_simple.py，输入以下代码（采用 tfrecords 方法读入，而不是直接读取 MNIST_data）。

```
#coding: utf-8
import tensorflow as tf
#import input_data
from read_and_decode import *
import pylab
import numpy as np
import os

#载入数据集
img, label=read_and_decode("./mnist_bmp_train/train.tfrecords")#训练数据集
```

```python
img_test, label_test=read_and_decode("./mnist_bmp_test/test.tfrecords")
#测试数据集

img_batch, label_batch=tf.train.shuffle_batch([img, label],
                                              batch_size=100, capacity=55000,
                                              min_after_dequeue=25000)
label_batch=tf.one_hot(label_batch,10,1,0)#'1,0'代表只有一个是1,其余为0;
'0,1'代表只有一个是0,其余为1

img_test_batch, label_test_batch=tf.train.shuffle_batch([img_test, label_test],
                                              batch_size=10000, capacity=20000,
                                              min_after_dequeue=10000)
label_test_batch=tf.one_hot(label_test_batch,10,1,0)
#定义两个placeholder
x=tf.placeholder(tf.float32,[None,784*1])
y_=tf.placeholder(tf.float32,[None,10])

#创建一个简单的神经网络
W=tf.Variable(tf.zeros([784*1,10]))
b=tf.Variable(tf.zeros([10]))
prediction=tf.nn.softmax(tf.matmul(x,W)+b)
#二次代价函数
loss=tf.reduce_mean(tf.square(y_-prediction))

#使用梯度下降法
train_step=tf.train.GradientDescentOptimizer(0.2).minimize(loss)

#初始化变量
init=tf.global_variables_initializer()

#结果存放在一个布尔型列表中, argmax 返回一维张量中最大的值所在的位置
correct_prediction=tf.equal(tf.argmax(y_,1),tf.argmax(prediction,1))

#求准确率
accuracy=tf.reduce_mean(tf.cast(correct_prediction,tf.float32))

coord=tf.train.Coordinator()
init=tf.global_variables_initializer()
with tf.Session() as sess:
    sess.run(init)
    #启动队列
    threads=tf.train.start_queue_runners(sess,coord=coord)
```

```python
for epoch in range(101):
    for batch in range(55):
        batch_xs, batch_ys=sess.run([img_batch, label_batch])
        sess.run(train_step, feed_dict={x:batch_xs.reshape(-1,784),y_:
        batch_ys})
    train_acc=sess.run(accuracy,feed_dict={x:batch_xs.reshape(-1,784),
    y_:batch_ys})
    print("Iter "+str(epoch)+",Training Accuracy "+str(train_acc))
batch_test_xs, batch_test_ys=sess.run([img_test_batch, label_test_batch])
test_acc=sess.run(accuracy,feed_dict={x:batch_test_xs.reshape(-1,784),
y_:batch_test_ys})
print("Testing Accuracy "+str(test_acc))

coord.request_stop()
coord.join(threads)
```

（2）运行 my_mnist_tfrecord_simple.py。

```
python my_mnist_tfrecord_simple.py
```

终端运行结果如图 3-16 所示。

```
Iter 80,Training Accuracy 0.91
Iter 81,Training Accuracy 0.87
Iter 82,Training Accuracy 0.93
Iter 83,Training Accuracy 0.92
Iter 84,Training Accuracy 0.94
Iter 85,Training Accuracy 0.88
Iter 86,Training Accuracy 0.9
Iter 87,Training Accuracy 0.93
Iter 88,Training Accuracy 0.94
Iter 89,Training Accuracy 0.91
Iter 90,Training Accuracy 0.91
Iter 91,Training Accuracy 0.88
Iter 92,Training Accuracy 0.92
Iter 93,Training Accuracy 0.92
Iter 94,Training Accuracy 0.9
Iter 95,Training Accuracy 0.92
Iter 96,Training Accuracy 0.91
Iter 97,Training Accuracy 0.92
Iter 98,Training Accuracy 0.91
Iter 99,Training Accuracy 0.93
Iter 100,Training Accuracy 0.89
Testing Accuracy 0.9056
```

图 3-16　Softmax Regression 神经网络模型的运行结果（tfrecord 数据）

3.4.2　卷积神经网络模型

（1）在/home/tensorflow/design/my_mnist_tfrecord/目录下新建空白文档并命名为 my_mnist_tfrecord.py，输入以下代码（采用 tfrecords 方法读入，而不是直接读取 MNIST_data）。

```python
#coding=utf-8
#---------------------
import tensorflow as tf
import matplotlib.pyplot as plt
from read_and_decode import *
import pylab
import numpy as np

img, label=read_and_decode("./mnist_bmp_train/train.tfrecords")
img_test, label_test=read_and_decode("./mnist_bmp_test/test.tfrecords")

img_batch, label_batch=tf.train.shuffle_batch([img, label],
                                    batch_size=100, capacity=55000,
                                    min_after_dequeue=25000)
label_batch=tf.one_hot(label_batch,10,1,0)#'1,0'代表只有一个是1,其余为0;
'0,1'代表只有一个是0,其余为1

img_test_batch, label_test_batch=tf.train.shuffle_batch([img_test, label_test],
                                    batch_size=10000, capacity=20000,
                                    min_after_dequeue=10000)
label_test_batch=tf.one_hot(label_test_batch,10,1,0)

#构建CNN网络结构
#自定义卷积函数
def conv2d(x,w):
    return tf.nn.conv2d(x,w,strides=[1,1,1,1],padding='SAME')
#自定义池化函数
def max_pool_2_2(x):
    return tf.nn.max_pool(x, ksize=[1,2,2,1], strides=[1,2,2,1], padding=
    'SAME')
#设置占位符,尺寸为样本输入和输出的尺寸t
x=tf.placeholder(tf.float32,[None,784])
y_=tf.placeholder(tf.float32,[None,10])
x_img=tf.reshape(x,[-1,28,28,1])

#设置第一个卷积层和池化层

w_conv1=tf.Variable(tf.truncated_normal([3,3,1,32],stddev=0.1),dtype=tf.
float32)
b_conv1=tf.Variable(tf.constant(0.1,shape=[32]))
h_conv1=tf.nn.relu(conv2d(x_img,w_conv1)+b_conv1)
h_pool1=max_pool_2_2(h_conv1)
```

```python
#设置第二个卷积层和池化层
w_conv2=tf.Variable(tf.truncated_normal([3,3,32,50],stddev=0.1))
b_conv2=tf.Variable(tf.constant(0.1,shape=[50]))
h_conv2=tf.nn.relu(conv2d(h_pool1,w_conv2)+b_conv2)
h_pool2=max_pool_2_2(h_conv2)
#设置第一个全连接层
w_fc1=tf.Variable(tf.truncated_normal([7*7*50,1024],stddev=0.1))
b_fc1=tf.Variable(tf.constant(0.1,shape=[1024]))
h_pool2_flat=tf.reshape(h_pool2,[-1,7*7*50])
h_fc1=tf.nn.relu(tf.matmul(h_pool2_flat,w_fc1)+b_fc1)

#dropout(权重失活)
h_fc1_drop=tf.nn.dropout(h_fc1,0.5)

#设置第二个全连接层
w_fc2=tf.Variable(tf.truncated_normal([1024,10],stddev=0.1))
b_fc2=tf.Variable(tf.constant(0.1,shape=[10]))
y_out=tf.nn.softmax(tf.matmul(h_fc1_drop,w_fc2)+b_fc2)

#使用交叉熵
loss=tf.reduce_mean(tf.nn.softmax_cross_entropy_with_logits(labels=y_,
logits=y_out))

#使用梯度下降法
train_step=tf.train.GradientDescentOptimizer(0.2).minimize(loss)

#建立正确率计算表达式
correct_prediction=tf.equal(tf.argmax(y_out,1),tf.argmax(y_,1))
accuracy=tf.reduce_mean(tf.cast(correct_prediction,tf.float32))

#开始读取数据,训练
coord=tf.train.Coordinator()
init=tf.global_variables_initializer()
with tf.Session() as sess:
    sess.run(init)
    #启动队列
    threads=tf.train.start_queue_runners(sess, coord=coord)
    for epoch in range(101):
        for batch in range(55):
            batch_xs, batch_ys=sess.run([img_batch, label_batch])
            sess.run(train_step,feed_dict={x:batch_xs.reshape(-1,784),
            y_:batch_ys})
            train_acc= sess.run(accuracy, feed_dict={x:batch_xs.reshape(-1,
            784),y_:batch_ys})
```

```
            print("Iter "+str(epoch)+",Training Accuracy "+str(train_acc))
    batch_test_xs, batch_test_ys=sess.run([img_test_batch, label_test_batch])
    test_acc=sess.run(accuracy,feed_dict={x:batch_test_xs.reshape(-1,
    784),y_:batch_test_ys})
    print("Testing Accuracy "+str(test_acc))
    coord.requet_stop()
    coord.join(threads)
```

（2）运行 my_mnist_tfrecord.py。

```
python my_mnist_tfrecord.py
```

终端运行结果如图 3-17 所示。

```
Iter 80,Training Accuracy 0.98
Iter 81,Training Accuracy 0.99
Iter 82,Training Accuracy 1.0
Iter 83,Training Accuracy 0.98
Iter 84,Training Accuracy 1.0
Iter 85,Training Accuracy 0.98
Iter 86,Training Accuracy 0.99
Iter 87,Training Accuracy 0.98
Iter 88,Training Accuracy 0.99
Iter 89,Training Accuracy 0.95
Iter 90,Training Accuracy 0.98
Iter 91,Training Accuracy 1.0
Iter 92,Training Accuracy 0.99
Iter 93,Training Accuracy 1.0
Iter 94,Training Accuracy 0.96
Iter 95,Training Accuracy 0.99
Iter 96,Training Accuracy 0.99
Iter 97,Training Accuracy 0.97
Iter 98,Training Accuracy 0.98
Iter 99,Training Accuracy 0.99
Iter 100,Training Accuracy 0.97
Testing Accuracy 0.9773
```

图 3-17 卷积神经网络模型的运行结果（tfrecord 数据）

习题 3

3.1 简述神经网络训练和测试两个阶段所完成的任务。
3.2 神经网络训练的目的是什么？
3.3 简述神经网络训练的过程。
3.4 全连接网络的特点是什么？
3.5 Loss 函数的作用是什么？
3.6 试写出交叉熵函数的数学表达式。
3.7 TensorFlow 中随机梯度下降函数是什么？
3.8 卷积神经网络的三个特征是什么？
3.9 局部感知的原理是什么？
3.10 简述 CNN 卷积运算的实现过程。

3.11 决定 CNN 中特征图大小的三个参数是什么？

3.12 CNN 中 padding 有哪几种方式？

3.13 试写出自定义 padding 尺寸与生成特征图尺寸之间的表达式。

3.14 简述 CNN 权值共享的原理。

3.15 简述 CNN 池化的原理及池化方式。

3.16 简述 ReLU 函数的原理，TensorFlow 中实现 ReLU 的 API 函数是什么？

第 4 章

OpenCL 基础

本章首先介绍 OpenCL 标准框架组成、基本概念、模型等基础知识，使读者对 OpenCL 有初步的认识和了解；然后给出一个设计实例，重点讲解 host 程序的编程步骤和 kernel 程序的编程要点。本章的学习将为编写可执行 host 代码和 kernel 代码奠定基础。

4.1 OpenCL 标准框架

OpenCL 是为面向异构系统进行并行运算而设计的行业标准框架。异构系统一般由 CPU、GPU 和其他设备构成，已经成为一类重要的计算平台。OpenCL 是直接满足这类异构系统需求的第一个行业标准。OpenCL 框架可以划分为 3 个主要组成部分：OpenCL 平台 API、OpenCL 运行 API 和 OpenCL 编程语言，如图 4-1 所示。其中，OpenCL 平台 API 和 OpenCL 运行 API 构成了宿主机（host）程序，OpenCL 编程语言对应内核（kernel）程序。一个完整的 OpenCL 编程程序包含 host 程序与 kernel 程序。

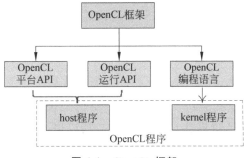

图 4-1 OpenCL 框架

1. OpenCL 平台 API

平台（platform）是宿主机和 OpenCL 框架管理下的若干设备（device）的组合。一个异构计算机上可以同时存在多个 OpenCL 平台。平台 API 定义了 host 程序发现

OpenCL 设备所用的函数以及这些函数的功能，另外还定义了为 OpenCL 应用程序创建上下文的函数。

例如，平台 API 可以查询哪些 OpenCL 设备可用，这些 OpenCL 设备具有什么特性等。在编写 host 程序时，通过调用相关平台 API 函数可以定义上下文。

2. OpenCL 运行 API

平台 API 为 OpenCL 应用程序定义上下文，运行 API 则定义了使用上下文的函数集。为了满足应用需求，运行 API 函数集庞大且复杂。

运行 API 的主要功能是管理上下文创建命令队列以及提交命令进入命令队列等运行时发生的所有操作。

运行 API 首先建立命令队列，并将命令队列关联到一个设备，一个上下文中可以同时存在多个命令队列。有了命令队列，就可以使用运行 API 定义内存对象和管理内存对象所需的其他对象。运行 API 还定义了与命令队列交互的命令，以及管理数据共享和对内核执行施加约束同步等命令。

3. OpenCL 编程语言

OpenCL 编程语言是指用来编写 kernel 函数的编程语言，是基于 ISO C99 标准的一个扩展子集，因此通常称为 OpenCL C 编程语言。

这个子集中去掉了 C99 的一些语言特性，包括递归函数、函数指针和位域。此外，该子集不支持完整的标准库集合，其中包含最常用的 stdio.h 和 stdlib.h。

此外，OpenCL 编程语言对 C99 进行了扩展，包括矢量类型及相关操作、地址控制限定符以及支持 OpenCL 应用的内核函数等。

4.2 OpenCL 基本概念基础

要理解 OpenCL 编程，首先需要理解一些基本概念。

1. 平台（Platform）

主机加上 OpenCL 框架管理下的若干设备构成了平台。通过平台，应用程序可以与设备共享资源并在设备上执行 kernel。实际使用中，基本上一个厂商对应一个平台，例如 Intel 和 AMD 公司都是这样。

2. 设备（Device）

设备是计算单元（compute units）的集合。举例来说，GPU、FPGA 是典型的设备。Intel 和 AMD 的多核 CPU 也提供 OpenCL 接口，所以也可以作为设备。

3. 上下文（Context）

上下文指 OpenCL 平台上共享和使用资源的环境，包括 kernel、device、memory objects、command queue 等。使用中一般一个平台对应一个上下文（context）。

4. 编程程序对象（Program Object）

程序对象包含内核程序代码和主机程序代码。针对 FPGA 平台，一般采用离线编译器将内核程序代码生成 FPGA 编程文件,将主机程序代码生成 host 可执行文件。

5. 内核函数(Kernel Function)

内核函数即实现内核功能的源程序,可以从主机端调用并运行在设备端。

6. 内存对象(Memory Object)

在主机和设备之间传递数据的对象,一般映射到 OpenCL 程序中的 global memory。有两种具体的类型:buffer object(缓存对象)和 image object(图像对象)。内存对象(memory object)一般与 kernel 函数的 global 参数相对应。

7. 指令队列(Command Queue)

主机与设备之间的交互通过指令完成,主机负责生成指令,并将指令放到一个队列中,从而构成指令队列。主机通过调度指令队列中的指令到指定的设备上执行。队列中指令的执行可以顺序,也可以乱序。一个设备可以对应多个指令队列。

4.3 OpenCL 程序的组成部分

如图 4-2 所示,OpenCL 程序划分成两部分:一部分是在设备上(如 CPU、GPU、FPGA 等)执行的 kernel 函数,对应图中的 Accelerator;另一部分是在主机(如 CPU、MCU 等)上运行的 host 程序,对应图中的 Host。在设备上执行的程序就是实现"异构"和"并行计算"的部分。为了能在设备上执行代码,程序员需要编写一个特殊的函数(kernel 函数)。这个函数需要使用 OpenCL C 语言编写。如前所述,OpenCL C 语言采用了 ISO C99 语言的一部分加上一些约束、关键字和数据类型。在主机上运行的程序通过 OpenCL 的 API 管理设备上运行的程序。主机程序的 API 用 C 语言编写,还有 C++、Java、Python 等高级语言接口。

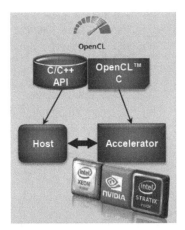

图 4-2 OpenCL 程序的划分

1. 内核函数

内核函数(kernel function)是在硬件加速器(OpenCL 设备)上运行的软件程序,通常用来计算密集型任务。图 4-3 为内核函数实现两个 n 维数组相加的例子。采用 OpenCL C 语言描述数学算法可得到 kernel 函数。

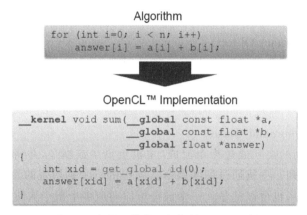

图 4-3 实现 n 维数组相加的 kernel 函数

2. 主程序

主程序(host program)是指在传统微处理器(一般指 CPU 或嵌入式微处理)上运行的软件程序。主机程序通过使用 host API 函数和 kernel 程序相互合作达到高效实现算法的目的。主机与 OpenCL 设备之间的交互就是使用一系列 API 函数库例程,这些例程定义了主处理器与内核的通信机制。如图 4-4 所示,函数 clEnqueueWriteBuffer() 实现数据从 host 到 FPGA 的复制,即 host 写 device。函数 clEnqueueReadBuffer() 实现 FPGA 到 host 的数据复制,即 host 读 device。函数 clEnqueueTask() 实现 kernel 函数的运行。

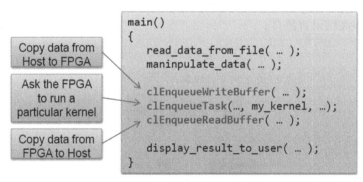

图 4-4 Host API 举例

4.4 OpenCL 框架的 4 种模型

OpenCL 支持大量不同类型的应用,笼统概括这些应用是很困难的,不过所有面向异构平台的应用都必须完成以下步骤:

(1) 发现构成异构系统的组件;

（2）查询所发现组件的特征，使软件能够适应不同硬件单元的特性；

（3）创建内核；

（4）建立并管理计算中设计的内存对象；

（5）在系统中正确的组件上按正确顺序执行内核；

（6）读取内核执行结果。

为了解释以上步骤的所有工作，引入了4种模型：平台模型、执行模型、内存模型和编程模型。

1. 平台模型

平台模型可以看作是异构系统的高层描述。一个 OpenCL 平台模型由一个主机和一个或多个设备构成。一个器件中包含多个计算单元 CU(compute unit)，一个计算单元 CU 中包含多个处理单元 PE(processing elements)，如图 4-5 所示。一个系统中可能存在多个平台，通常平台彼此之间不进行交互。一个平台对应一家厂商提供的 SDK。图 4-6 为 Intel FPGA 的一款 OpenCL 平台，包括主机和 FPGA。

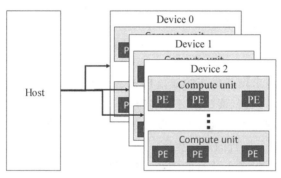

图 4-5　平台模型

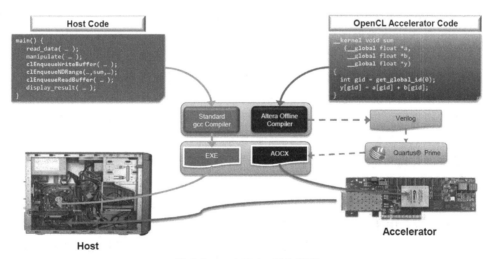

图 4-6　Intel FPGA 平台模型

2. 执行模型

执行模型是指令流在异构平台上执行的抽象表示。OpenCL 的执行模型可以分为两部分，一部分是在 host 上执行的主程序，另一部分是在 OpenCL 设备上执行的内核程序，OpenCL 通过主程序定义上下文并管理内核程序在 OpenCL 设备上的执行。

执行模型定义主机与设备之间如何通信、主机如何及何时在设备上执行程序，如图 4-7 所示。使用一个或多个设备创建 OpenCL 上下文，OpenCL 上下文为主机-设备交互、内存管理、设备控制等提供了一个环境。

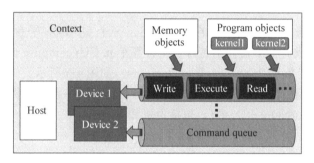

图 4-7 执行模型

命令队列是主机请求设备操作的通信机制。每个设备必须至少关联一个命令队列。主机发送的命令包括写数据、读数据、执行内核等。这些命令被存储在命令队列中并适当地发出。主机负责管理命令的执行顺序以及命令之间的依赖关系。

(1) OpenCL 设备上执行的内核程序。

内核是在 OpenCL 设备上执行的函数。执行内核的单元称为工作节点(work-item)。若干 work-item 组成一个工作组(work-group)。整个 work-item 的集合称为 NDRange，其中，work-group 和 work-item 可分为 N 个维度，其中 N 的最大值为 3。NDRange 和 work-group 的大小由主程序指定。

在 host 创建一个 kernel 程序之前，必须先为该 kernel 创建一个标识了索引的工作空间，kernel 会在工作空间中的每个节点上执行。该工作空间可以是一维、二维或者三维的。每个工作节点相应维度上的索引都被定义为节点在该维度上的全局 ID(global ID)。所有工作节点都将执行相同的 kernel 程序，但是由于路径不同，计算出来的数据可能是不同的。OpenCL 在对工作空间提供了全局索引之外，也提供了较小粒度的工作组空间。工作组的维度必须和整个工作空间的维度相同，并且整个工作空间在每个维度上都必须能等分成若干个工作区间。每个工作组都有一个唯一的工作组 ID(work group ID)，工作组内部的节点相对该工作组的位置索引被称为局部 ID(local ID)。

OpenCL 规范中使用 NDRange 定义 N-维索引空间，它由一个长度为 N 的整数数组构成(N 为 1、2 或 3，对应一维、二维或三维空间)，该数组中的元素指定了工作空间每个维度上工作节点的个数。工作节点的 global ID 和 local ID 以及 work group ID 都是 N 维的，每个维度都从 0 开始依次加 1。如图 4-8 所示，get_global_id(0)的作用是获取第一维空间的全局 ID，get_local_id(0)的作用是获取第一维空间中一个 work group 中 work item 的局部 ID,get_group_id(0)的作用是获取第一维空间中 work group 的 ID。

NDRange												
get_global_id(0)	0	1	2	3	4	5	6	7	8	9	10	11
get_local_id(0)	0	1	2	3	0	1	2	3	0	1	2	3
get_group_id(0)	0	0	0	0	1	1	1	1	2	2	2	2

图 4-8 NDRange 举例

（2）OpenCL 通过上下文管理整个平台的运行。

上下文中管理的资源包括运行平台上的所有 OpenCL 设备,运行在设备上的 kernel 函数对象,以及 host 和 OpenCL 设备的交互内存对象。用户可以使用 OpenCL API 创建和使用上下文,host 进而利用上下文建立一个或多个命令队列（command queue）以协调某个 OpenCL 设备上内核的执行。host 会通过 API 将各个将要在 OpenCL 设备上执行的命令放入 command queue 中等待调度。这些命令可以是执行 kernel 的命令,也可以是读写内存对象的命令或者同步命令等。由于这些命令是在 host 上进行调度而在 OpenCL 设备上执行的,因此命令的执行完全是异步的,OpenCL 为 command queue 中的多个异步程序提供了两种执行模式：一种为顺序执行,即按照命令放入队列的顺序依次执行,只有前面的命令全部结束之后才会启动后面的命令；另一种为乱序执行,即后续命令不必等待前面的命令运行完毕即可执行,在此模式下,用户必须通过同步命令管理各个命令之间的执行顺序。用户可以为 command queue 中执行的每个命令指定一个事件对象以控制和协调该命令的执行。同一个上下文可能会关联不同 OpenCL 设备之间的 command queues 或者同一个 OpenCL 设备上的多个 command queues,OpenCL 规范没有对 command queues 之间的同步做任何显式规定。

3. 内存模型

主机和设备中的存储器分别称为主机存储器和设备存储器。主机存储器只能由主机访问。kernel 程序运行在设备上,对应设备存储器。OpenCL 将 kernel 程序中用到的设备存储器分为四类,如图 4-9 所示。

（1）全局内存。

为了让设备使用主机存储器中的数据,首先应将这些数据传输到设备的全局内存（global memory）。全局内存可以被主机和设备访问。工作空间内的所有工作节点都可以读写此类内存中的任意元素。OpenCL C 提供了缓存 global buffer 的内建函数。

（2）常量内存。

常量内存（constant memory）是设备的只读内存。工作空间内的所有工作节点可以只读此类内存中的任意元素。host 负责分配和初始化 constant buffer,它在内核执行过程中保持不变。

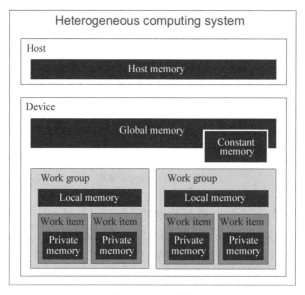

图 4-9　内存模型

（3）本地内存。

本地内存（local memory）从属于一个工作组的内存，同一个工作组中的所有工作节点都可以共享使用该类内存，而其他工作组无法访问这些数据。

（4）私有内存。

私有内存（private memory）只从属于当前的工作节点。对于一个工作节点内部的 private buffer，其他节点是完全不可见的。每个工作节点都可以有自己的专用内存，其他工作节点无法访问它。

设备存储器的分配及访问权限如表 4-1 所示。

表 4-1　设备存储器的分配及访问权限

		全局	常量	本地	私有
host 端	分配	动态	动态	动态	不可分配
	访问	可读写	可读写	不可访问	不可访问
kernel 端	分配	不可分配	静态	静态	静态
	访问	可读写	只读	可读写	可读写

4. 编程模型

OpenCL 支持按数据并行的编程模型和按任务并行的编程模型，同时也支持这两种模型混合而成的混合编程模型。其中，按数据并行的模型在编写 OpenCL 并行程序中最为经典。

数据并行模型是指同一系列的指令作用在内存对象的不同元素上，即在不同内存元素上按这个指令序列定义了统一的运算。用户可以利用工作节点的 global ID 或 local ID

映射该工作节点所作用的内存元素。在一个严格意义上的数据并行编程模型中,每个工作节点和该节点上 kernel 程序所作用的内存之间都有着一对一的映射关系。OpenCL 实现的是一个宽松的数据并行模型,并不要求一定要有严格的一对一映射。OpenCL 提供了一种分级数据并行编程模型,其分级方式有两种:显式分级模型要求用户不仅要规定用于数据并行计算的所有工作节点的数目,同时也必须规定每个节点所属的工作组;隐式分级模型将后者交由 OpenCL 实现管理。

任务并行编程模型是指工作空间内的每个工作节点在执行 kernel 程序时相对于其他节点是绝对独立的。在这种模式下,每个工作节点都相当于工作在一个单一的计算单元内,该单元内只有单一工作组,该工作组中只有该节点本身在执行。用户可以通过以下方法实现按任务并行:

① 使用 OpenCL 设备支持的向量类型数据结构;

② 同时执行或选择性执行多个 kernels;

③ 在执行 kernels 的同时交叉执行一些 native kernels 程序。

任何多线程并行计算模型都可能涉及同步问题。OpenCL 提供了以下两个领域的同步:

① 在同一个工作组中所有工作节点之间的同步;

单个工作组内的所有工作节点之间的同步是通过工作组的阻断函数实现的,而工作组之间没有相应的阻断函数,是无法动态同步的。

② 同一个上下文中不同 command queues 之间和同一个 command queue 的不同 commands 之间的同步。

OpenCL 也为 command queue 提供了类似的阻断函数,以保证该 queue 中的 command 在阻断之前全部完成,这种阻断对读写内存对象的命令也适用。该阻断仅用于 command queue 内部。OpenCL 没有提供特殊的 API 同步不同的 command queue,但用户可以利用每个 command 所关联的 event 同步各个 command,进而达到同步 command queue 的目的。

4.5 编写第一个 OpenCL 程序

从上面对 OpenCL 程序的描述可知,一个完整的 OpenCL 程序由两部分代码组成,一是 host 程序,二是 kernel 程序。host 程序是在 CPU 或者嵌入式 MCU(如 ARM 处理器等)中运行的。host 程序采用标准 C/C++语言编写,用 C/C++编译器进行编译,在 Linux 环境下一般采用 gcc 编译器。kernel 程序在 GPU 或 FPGA 等加速设备上运行,采用 OpenCL C 语言编写。对于 FPGA 设备,为了提供 FPGA 的编程数据,kernel 程序结合 BSP 一起采用嵌入在 SDK 中的离线编译器进行编译。

4.5.1 kernel 程序

本例程的 kernel 程序实现的功能为:对一个长度为 100 的数组中的所有元素进行乘以 2 的操作。

kernel 代码如下。

```
__kernel void mul_2 (__global const float * restrict dev_x,
                     __global float * restrict dev_y)
{
    for (int k=0;k< 100;k++)
    {
        dev_y[k]=dev_x[k] * 2;
    }
}
```

kernel 程序以关键词 __kernel 开头,返回类型为 void。mul_2 为该 kernel 的名称。__global 为地址空间限定符,const 一般与地址空间限定符一起使用,代表只读存储空间。float 代表数据类型,restrict 代表限定的指针,不会出现多个指针指向同一块内存的情况,与 register 关键字类似,这个也是提供给编译器优化的,保证只有一个指针会指向这块内存,编译器能更高效地进行一些处理而不用担心影响其他指针。dev_x 为输入参数,dev_y 为输出参数。前缀 dev_表示在设备上的数据。for 循环语句实现数组内元素的乘 2 运算。

这种代码编写方式称为 single work item 方式,与之相对应的是称为 NDRange 的方式。

4.5.2 host 程序

1. host 编程步骤

host 程序的编程步骤如图 4-10 所示。下面结合本例程实现的功能介绍各个步骤的实现方法及相关知识点。

(1) 获取 OpenCL 平台和设备信息。

```
//Get the OpenCL platform.
  clGetPlatformIDs(1, &platform, NULL);
//  Obtain the available OpenCL devices.
  clGetDeviceIDs(platform, CL_DEVICE_TYPE_ALL, 1,&device,NULL);
```

特定型号的开发板及 BSP 获取 OpenCL 平台和设备信息的方式是不变的,编写应用代码时可直接复制使用,无须修改。

(2) 创建上下文和命令队列。

```
//Create the context.
  context=clCreateContext(NULL, 1, &device, NULL, NULL, &status);

//Create the command queue.
  queue=clCreateCommandQueue(context, device, CL_QUEUE_PROFILING_ENABLE,
&status);
```

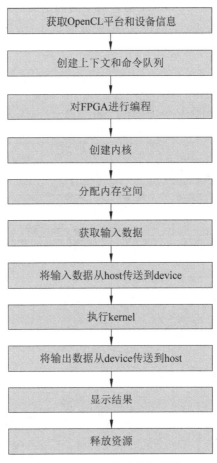

图 4-10 host 程序编程步骤

(3) 对 FPGA 进行编程。

```
//Create the program.
  std::string binary_file=getBoardBinaryFile(aocx_file, device);
  program=createProgramFromBinary(context, binary_file.c_str(), &device,
1);

//Build the program that was just created.
  status=clBuildProgram(program, 0, NULL, "", NULL, NULL);
```

(4) 创建内核。

```
//create the kernel
my_first_kernel=clCreateKernel(program, kernel_name, &status);
```

(5) 分配内存空间。

内存空间的分配分为两种类型。一种是在 host 主机上分配内存空间,用来存储输入/输出数据;另一种是为了访问 device 的存储空间而定义的内存对象(memory object),内存对象定义了该内存在 device 上的位置,用关键词 cl_mem 进行声明,使用 clCreatBuffer 函数定义。

```
//allocate and initialize the data vectors of host side
    cl_float * x, * y;
    x=(cl_float *)alignedMalloc(sizeof(cl_float) * x_size);
    y=(cl_float *)alignedMalloc(sizeof(cl_float) * y_size);

//create the data buffer of device side
    cl_mem dev_x, dev_y;
    dev_x=clCreateBuffer(context, CL_MEM_READ_WRITE, sizeof(cl_float) * x
_size, NULL, &status);
    dev_y=clCreateBuffer(context, CL_MEM_READ_WRITE, sizeof(cl_float) * y
_size, NULL, &status);
```

(6) 获取输入数据。

```
//load input data
    for(int i=0;i< 100;i++)
        {x[i]=i;}
```

(7) 将数据从 host 传送到 device。

```
//Transfer data from host side to device side
    status=clEnqueueWriteBuffer(queue, dev_x, CL_TRUE, 0, sizeof(cl_float) *
x_size, x, 0, NULL, NULL);
```

(8) 执行 kernel。

为了执行 kernel,kernel 函数的参数需要指定如下。

```
//set the kernel arguments
    status=clSetKernelArg(my_first_kernel,0, sizeof(cl_mem), (void*)&dev_x);
    status=clSetKernelArg(my_first_kernel,1, sizeof(cl_mem), (void*)&dev_y);

    cl_event event_kernel;
    static const size_t GSize[]={1};
    static const size_t WSize[]={1};

//launch kernel
    status=clEnqueueNDRangeKernel(queue, my_first_kernel, 1,0, GSize, WSize,
0, NULL, &event_kernel);
```

(9) 将输出数据从 device 传送到 host。

```
//Transfer data from  device side to host side
    status=clEnqueueReadBuffer(queue, dev_y, CL_TRUE, 0, sizeof(cl_float)
* y_size, y, 0, NULL, NULL);
```

（10）显示结果。

```
//display result
    for(int i=0;i< y_size;i++)
       {
            printf("i=%i,",i);
            printf("x=%.1f,",x[i]);
    printf("y=%.1f\n",y[i]);
       }
```

（11）释放资源。

```
//release all resource
  clFlush(queue);
  clFinish(queue);
  //device side
  clReleaseMemObject(dev_x);
  clReleaseMemObject(dev_y);
  clReleaseKernel(my_first_kernel);
  clReleaseProgram(program);
  clReleaseCommandQueue(queue);
  clReleaseContext(context);
  //hose side
  free(x);
  free(y);
```

2. 完整的 host 程序代码

```
//声明头文件及名称空间
#include < assert.h>
#include < stdio.h>
#include < stdlib.h>
#include < math.h>
#include < cstring>
#include "CL/opencl.h"
#include "AOCLUtils/aocl_utils.h"
using namespace aocl_utils;
using namespace std;
/***在编写新 host 代码时,以上内容无须修改******/
//OpenCL runtime configuration
static cl_platform_id platform=NULL;        //定义平台名称
static cl_device_id device=NULL;            //定义设备名称
static cl_context context=NULL;             //定义上下文名称
static cl_command_queue queue=NULL;         //定义队列名称
```

```
static cl_kernel my_first_kernel=NULL;        //定义内核名称
static cl_program program=NULL;               //定义编程对象名称

//define constant
#define x_size 100
#define y_size 100

//define kernel file
const char * source_file="mul_2.cl";          //kernel 文件名称为 mul_2.cl
const char * kernel_name="mul_2";   //kernel 代码中 kernel 函数名称
                                    //一个 kernel 代码中可以包含多个 kernel 函数
//define fpga program fiel
const char * aocx_file="mul_2";      //定义 FPGA 编程文件 aocx 文件名

cl_int status;

int main() {
  //Get the OpenCL platform
  clGetPlatformIDs(1, &platform, NULL);
  //Obtain the available OpenCL devices.
  clGetDeviceIDs(platform, CL_DEVICE_TYPE_ALL, 1, &device, NULL);
  //Create the context
  context=clCreateContext(NULL, 1, &device, NULL, NULL, &status);
  //Create the command queue
  queue=clCreateCommandQueue(context, device, CL_QUEUE_PROFILING_ENABLE,
&status);
  //Create the program
  std::string binary_file=getBoardBinaryFile(aocx_file, device);
  program=createProgramFromBinary(context, binary_file.c_str(), &device, 1);
  //Build the program that was just created
    status=clBuildProgram(program, 0, NULL, "", NULL, NULL);
  //create the kernel
    my_first_kernel=clCreateKernel(program, kernel_name, &status);
  //allocate and initialize the data vectors of host side
    cl_float * x, * y;
    x=(cl_float *)alignedMalloc(sizeof(cl_float) * x_size);//
    y=(cl_float *)alignedMalloc(sizeof(cl_float) * y_size);//
  //create the data buffer of device side
    cl_mem dev_x, dev_y;
    dev_x=clCreateBuffer(context, CL_MEM_READ_WRITE, sizeof(cl_float) * x_
size, NULL, &status);

    dev_y=clCreateBuffer(context, CL_MEM_READ_WRITE, sizeof(cl_float) * y_
size, NULL, &status);
```

```c
//load input data
   for(int i=0;i< 100;i++)
     {x[i]=i;}
//Transfer data from host side to device side
   status=clEnqueueWriteBuffer(queue, dev_x, CL_TRUE, 0, sizeof(cl_float) *
x_size, x, 0, NULL, NULL);
//set the kernel arguments
  status=clSetKernelArg(my_first_kernel,0, sizeof(cl_mem), (void*)&dev_x);
  status=clSetKernelArg(my_first_kernel,1, sizeof(cl_mem), (void*)&dev_y);
//launch kernel
   cl_event event_kernel;
   static const size_t GSize[]={1};
   static const size_t WSize[]={1};
   status=clEnqueueNDRangeKernel(queue, my_first_kernel, 1,0, GSize, WSize, 0,
   NULL, &event_kernel);
//read the output
   status=clEnqueueReadBuffer(queue, dev_y, CL_TRUE, 0, sizeof(cl_float)
   * y_size, y, 0, NULL, NULL);
//display result
     for(int i=0;i< y_size;i++)
        {
           printf("i=%i,",i);
           printf("x=%.1f,",x[i]);
         printf("y=%.1f\n",y[i]);
         }
//release all resource
   clFlush(queue);
   clFinish(queue);
   //device side
   clReleaseMemObject(dev_x);
   clReleaseMemObject(dev_y);
   clReleaseKernel(my_first_kernel);
   clReleaseProgram(program);
   clReleaseCommandQueue(queue);
   clReleaseContext(context);
   //hose side
   free(x);
   free(y);
  return 0;
}
//cleanup()函数在本书中为一个空函数,为了配合开发板提供的bsp头文件,需要定义该函
数,否则会报错
void cleanup()
{

}
```

3. 执行结果

将 kernel 程序编译为 aocx 文件,将 host 程序编译为可执行文件,DE10_nano 开发板上的执行结果如图 4-11 所示。

```
i=0,x=0.0,y=0.0
i=1,x=1.0,y=2.0
i=2,x=2.0,y=4.0
i=3,x=3.0,y=6.0
i=4,x=4.0,y=8.0
i=5,x=5.0,y=10.0
i=6,x=6.0,y=12.0
i=7,x=7.0,y=14.0
i=8,x=8.0,y=16.0
i=9,x=9.0,y=18.0
i=10,x=10.0,y=20.0
i=11,x=11.0,y=22.0
i=12,x=12.0,y=24.0
i=13,x=13.0,y=26.0
i=14,x=14.0,y=28.0
i=15,x=15.0,y=30.0
i=16,x=16.0,y=32.0
i=17,x=17.0,y=34.0
i=18,x=18.0,y=36.0
i=19,x=19.0,y=38.0
i=20,x=20.0,y=40.0
...
i=90,x=90.0,y=180.0
i=91,x=91.0,y=182.0
i=92,x=92.0,y=184.0
i=93,x=93.0,y=186.0
i=94,x=94.0,y=188.0
i=95,x=95.0,y=190.0
i=96,x=96.0,y=192.0
i=97,x=97.0,y=194.0
i=98,x=98.0,y=196.0
i=99,x=99.0,y=198.0
```

图 4-11 执行结果

4.6 OpenCL 基本知识点

4.6.1 kernel 函数格式

kernel 函数的编写格式如下:

```
__kernel void 函数名(输入参数列表,输出参数列表);
{
    功能定义;
}
```

kernel 函数以关键词__kernel 开头,返回类型为 void。OpenCL C 比 C 多出了一些函数限定符,在本例中用到了__kernel(以两个下画线开始,也可简写为 kernel),这也是所有 OpenCL C 中必须用到的。函数加注__kernel 表示其只能运行于 OpenCL 设备上,并且该函数可以被 host 调用,kernel 中的其他函数也可以像调用普通函数一样调用含有该前缀的函数。

4.6.2 kernel 编程模式

kernel 编程模式有两种,分别为 single work item 和 Ndrange。若代码中没有使用任何工作项的函数,如 get_local_id()和 get_global_id(),则编程模式为 single work item 模式,否则为 NDRange 模式。4.4.1 节中的 kernel 代码采用的是 single work item 模式,若采用 NDRange 模式,则 kernel 代码将变为如下代码。

```
__kernel void mul_2 (__global const float * restrict dev_x,
                    __global float * restrict dev_y)
{
    int k=get_global_id(0);
        dev_y[k]=dev_x[k] * 2;
}
```

注:get_global_id(0)实现获取 work item 在 NDRange 的第一维空间中的 ID。

为了支持不同的 kernel 程序编程模式,host 程序需要根据不同的编程模式进行 NDRange 参数的设置。

(1) single work item 模式,host 程序对应部分代码。

```
static const size_t GSize[]={1};
static const size_t WSize[]={1};
status=clEnqueueNDRangeKernel(queue, my_first_kernel, 1,0, GSize, WSize, 0,
NULL, &event_kernel);
```

(2) NDRange 模式,host 程序对应部分代码。

```
static const size_t GSize[]={100};
static const size_t WSize[]={1};
status=clEnqueueNDRangeKernel(queue, my_first_kernel, 1,0, GSize, WSize, 0,
NULL, &event_kernel);
```

4.6.3 kernel 地址限定符

OpenCL C 增加了一些地址空间限定符,包括__global、__local、__const 和__private。这些地址限定符可以用来声明变量,以表明该变量对象所使用的内存区域。以上述地址空间限定符声明的对象会被分配到指定的地址空间,没有地址空间限定符声明的对象会被分配到通用地址空间。

__global 或简写为 global,表示该声明内存对象使用的是从全局内存空间分配出来的内存,工作空间内的所有工作节点都可以读/写这块被声明的内存。

__const 或简写为 const,与 global 相同,它使用从全局内存空间分配出来的内存,可以在所有工作节点做只读访问。在 OpenCL C 中试图对 constant 对象进行写操作会产生编译错误。

＿_local 或简写为 local，声明内存对象使用的是 local memory pool 中的内存，仅能在同一个工作组的不同工作节点之间访问。

＿_private 或简写为 private，所有在 OpenCL C 函数内部使用的变量或者传入函数内部的参数都是 private 类型，用户在声明 private 类型的变量时可以省略此限定符（因此无地址空间限定符的变量都属于此类型）。private 类型的对象仅可在其所在的工作节点内部使用。

4.6.4 kernel 语句描述

可以使用 C99 语言的语句：
① 操作符，如＋，－，＊，％，＜＜，?：，&，&&，～，!，＋＋，＝＝。
② 数学函数，如 sin,acos,log,exp,pow,floor,fabs,fma,fmod。
③ 调用用户定义的非 kernel 函数。
④ 使用数据流控制声明，如 if 语句，for 语句等。
⑤ 使用 C99 预编译类型，如♯include 等。

4.6.5 kernel 数据类型

① 标量数据类型。包括 char,ushort,int,uint,long,float,double,bool 等。在 host 端，建议使用 cl_前缀以确保数据长度匹配和最大可移植性。
② 图像类型。包括 image2d_t,image3d_t,sampler_t。
③ 用户自定义结构体。
④ 向量数据类型。向量的维数取值：2,3,4,8,16。host 程序和 kernel 代码中都可以使用，按照向量长度对齐。例如：char2,ushort3,int8,float16（kernel 函数格式），cl_char2,cl_ushort3,cl_int8,cl_float16（host 程序格式）。

4.6.6 kernel 编程限制

① 没有指向函数的指针。
② 没有递归调用。
③ 没有预定义的标识符。
④ 没有可写的静态变量。

习题 4

4.1 OpenCL 框架的三个主要组成部分是什么？
4.2 一个完整的 OpenCL 程序包含哪两个组成部分？
4.3 简述 OpenCL 平台、设备、上下文、编程文件、核函数等基本概念。
4.4 主机与设备之间采用什么 API 函数实现数据交互？

4.5　OpenCL 框架的四种模型是什么？
4.6　什么是 work-item？什么是 work-group？
4.7　什么是 NDRange？N 的取值范围是什么？
4.8　简述 global ID、work-group ID、local ID 三者之间的关系。
4.9　kernel 程序使用的存储器可以分为哪四种类型？各有什么特点？
4.10　kernel 程序代码的编程方式主要有哪两种？如何区分这两种编程方式？
4.11　简述 kernel 程序的语法格式。
4.12　简述 host 程序的编程步骤。
4.13　kernel 程序中的地址限定符有哪些？
4.14　kernel 程序支持的数据类型有哪些？
4.15　kernel 程序的编程限制有哪些？

第 5 章

面向 Intel FPGA 的 OpenCL 运行平台搭建

本章介绍在 Ubuntu 环境下搭建 OpenCL 运行平台的方法,主要包括 Quartus prime、OpenCL SDK、EDS 等软件的安装,DE10_nano BSP 的安装及系统环境变量的设置,SD 卡 img 文件的烧写,minicom 驱动的安装与参数设置,开发板与 PC 通过以太网实现数据传输的方法等。通过本章的学习,读者可以掌握面向 Intel FPGA 的 OpenCL 运行平台的搭建方法。

5.1 搭建 OpenCL 平台的软硬件要求

本书使用的 OpenCL 运行平台需要以下软硬件部分:
(1) 友晶科技的 DE10_nano 开发板;
(2) microSD 卡(至少 4GB,开发板自带);
(3) microSD 读卡器;
(4) USB 线(A to mini-B,开发板自带);
(5) 以太网电缆及网络路由器;
(6) 具备以下条件的 PC。
- USB 接口。
- 32GB 存储器(至少 16GB)。
- 64 位操作系统(Ubuntu 16.04 LTS)。
- win32 Disk Imager(运行在 Windows 环境下)。
- minicom 软件。
- Intel Quartus Prime 软件。
- Intel OpenCL 软件。
- Intel SoC EDS 软件。
- DE10_nano BSP。

5.2 面向 OpenCL 应用的 DE10_nano 开发板简介

DE10_nano 开发板是一个基于 Intel FPGA 的片上系统(system on chip,SoC)硬件设计平台,该平台结合了嵌入式 Cortex-A9 双核处理器和可编程逻辑,具有很高的设计灵活性。设计平台的资源包括:具有 110k 个可编程逻辑单元的 FPGA,型号为 CycloneV 5CSEBA6U23I7;时钟频率可达 800MHz 的双核 ARM Cortex-A9 处理器;32 位 1GB DDR3 SDRAM;1 Gigabit Ethernet PHY 等。其中,以双核 ARM 处理器、外设和存储器等组成了一个硬处理器系统(hard processor system,HPS)。

针对基于 DE10_nano 开发板的 OpenCL 开发设计需要注意以下几个问题。

(1) FPGA 的配置方式由 HPS 系统完成,HPS 系统执行存储在 SD 卡上的镜像文件。

(2) FPGA 配置方式的开关设置如图 5-1 所示,应将 MSEL[4..0]设置为 01010。开关拨到 ON 端为 0,拨到另一端为 1。

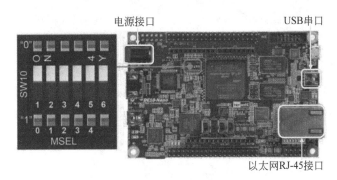

图 5-1 DE10_nano 开发板 MSEL 开关设置

(3) 开发板运行 OpenCL 程序需要连接三个接口:一是电源接口,二是 USB 串口,三是 RJ-45 网线接口。三个接口的位置在图 5-1 中进行了标识。电源接口为开发板供电,USB 串口用来实现 PC 与开发板之间的信息交互,RJ-45 网线实现 PC 与开发板之间的数据复制。

(4) 图 5-2 为 DE10_nano 开发板的 OpenCL 架构。整个 OpenCL 程序由 kernel 程序和 host 程序组成。kernel 程序由 PC 操作系统中安装的 DE10_nano 开发板的 BSP 进行编译以生成 aocx 文件,在 FPGA 可编程逻辑上运行。host 程序同样由 PC 操作系统中安装的 DE10_nano 开发板的 BSP 进行编译以生成可执行文件,由 ARM 处理器执行。

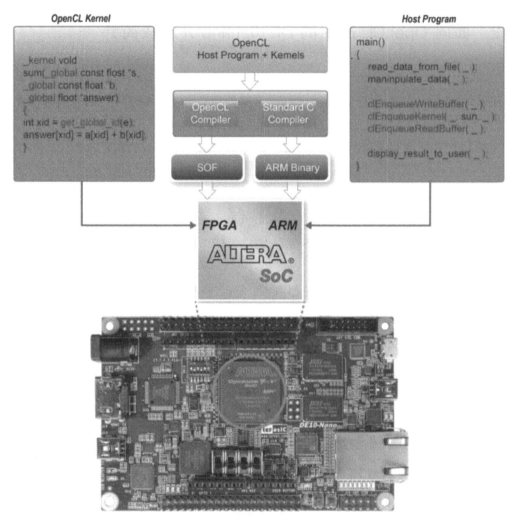

图 5-2　DE10_nano 开发板的 OpenCL 架构

5.3　平台所需软件下载

5.3.1　Quartus Prime Standard 下载

本书选用的软件版本为 18.1，下载链接为：http://fpgasoftware.intel.com/18.1/?edition=standard&platform=linux&download_manager=dlm3。

选择版本类型：Standard。

选择版本：18.1。

操作系统：Linux。

下载方法：方法(1)和方法(2)任选其一。

(1) 在组合文件标签下的目录中选择下载 Complete Download 中的 Quartus-18.1.0.

625-linux-complete.tar,如图 5-3 所示。

图 5-3   Quartus Prime Standard Complete 下载资源网站截图

（2）在独立文件标签下的目录中选择下载 Quartus Prime Standard Edition 中的 Quartus Prime (includes Nios II EDS)，以及 Devices 中的 Cyclone V device support，如图 5-4 所示。

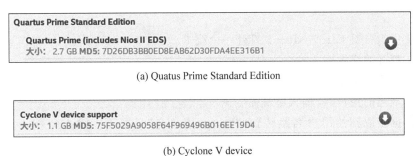

图 5-4   独立下载资源网站截图

单击下载按钮后，若没有登录或没有注册过账号，则网址会提示登录或注册，按照网站提示登录或注册即可。

5.3.2   Intel FPGA SDK for OpenCL 下载

在独立文件标签中选择下载 Intel FPGA SDK for OpenCL，如图 5-5 所示。

图 5-5   Intel FPGA SDK for OpenCL 下载资源网站截图

5.3.3   Intel SoC FPGA EDS 下载

在独立文件标签中选择下载 Intel SoC FPGA Embedded Development Suite Standard Edition，如图 5-6 所示。

图 5-6   Intel SoC FPGA EDS 下载资源网站截图

5.4 平台所需软件安装

5.4.1 安装 Quartus Prime Standard Edition+ Intel FPGA SDK for OpenCL

1. 安装步骤

在 Ubuntu 系统中,使用 firefox 网络浏览器的默认下载路径为"/home/用户名/下载"。注意:这里的用户名为 Ubuntu 系统当前用户的名字,作者的用户名为 ubuntu602,因此默认下载路径为"/home/ubuntu602/下载"。

按以下步骤进行操作,以 ubuntu602 作为用户名为例进行说明,读者在具体操作时,需要将 ubuntu602 替换为自己 Ubuntu 系统的当前用户名。获取当前用户名的命令为 who。

(1) 使用组合键 Ctrl+Alt+T 打开一个终端。
(2) 在打开的终端输入命令 sudo su。
(3) 输入管理员密码,获得管理员权限。
(4) 获取当前用户名,输入终端命令 who。
(5) 根据获取到的用户名跳转到"/home/用户名"目录下。输入命令 cd /home/用户名。

注意:用户名要用(4)中获得的真实用户名替换,作者的用户名为 ubuntu602,因此输入的命令为: cd /home/ubuntu602。

(6) 跳转到"下载"目录下,输入终端命令 cd 下载。
(7) 确认安装文件存在于"下载"目录下,输入终端命令 ls。
(8) 赋予安装文件可执行权限,输入终端命令 chmod +x QuartusSetup-18.1.0.625-linux.run AOCLSetup-18.1.0.625-linux.run。
(9) 执行安装,输入终端命令 ./QuartusSetup-18.1.0.625-linux.run。

图 5-7 为作者 PC 端安装过程中输入命令的截图。

```
ubuntu602@ubuntu602:~$ sudo su
[sudo] ubuntu602 的密码:
root@ubuntu602:/home/ubuntu602# who
ubuntu602 tty7         2019-01-29 09:40 (:0)
root@ubuntu602:/home/ubuntu602# cd /home/ubuntu602
root@ubuntu602:/home/ubuntu602# cd 下载
root@ubuntu602:/home/ubuntu602/下载# ls
AOCLSetup-18.1.0.625-linux.run         t10k-images-idx3-ubyte.gz
cyclonev-18.1.0.625.qdz                t10k-labels-idx1-ubyte.gz
NVIDIA-Linux-x86_64-410.93.run         train-images-idx3-ubyte.gz
QuartusSetup-18.1.0.625-linux.run      train-labels-idx1-ubyte.gz
SoCEDSSetup-18.1.0.625-linux.run
root@ubuntu602:/home/ubuntu602/下载# chmod +x QuartusSetup-18.1.0.625-linux.run
 AOCLSetup-18.1.0.625-linux.run
root@ubuntu602:/home/ubuntu602/下载# ./QuartusSetup-18.1.0.625-linux.run
```

图 5-7 输入命令截图

安装过程中出现如图 5-8(a)~图 5-8(h)所示的 8 个对话框。图 5-8(b)中需要选择"I accept the agreement"选项。

注意：图 5-8(c)中的安装路径默认为"/root/intelFPGA/18.1"，因为"/root"目录采用图形界面访问不方便，因此需要将安装路径修改为"/home/intelFPGA/18.1"，如图 5-8(d)所示。如果图 5-8(e)中的信息不全，则说明相关软件没有下载到"下载"目录下，这时单击图中的 Cancel 按钮即可取消安装，完成相应软件的下载后，重新开始安装。在图 5-8(h)中只保留 Create shortcuts on Desktop 选项，然后单击 Finish 按钮完成安装。其余对话框保持默认设置即可。

(a) 对话框1

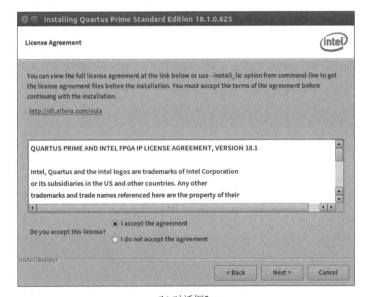

(b) 对话框2

图 5-8 安装过程中出现的对话框

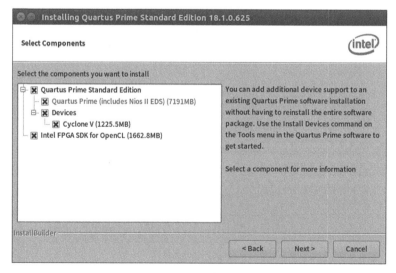

(c) 对话框3

(d) 对话框4

(e) 对话框5

图 5-8 （续）

(f) 对话框6

(g) 对话框7

(h) 对话框8

图 5-8 （续）

若桌面快捷方式无法创建,则需要在"/home/intelFPGA/18.1/quartus/bin"目录下运行终端命令 sudo ./quartus。

2. License 配置

切换到"/home/intelFPGA/18.1/quartus/bin"目录下,运行命令 sudo ./quartus。Quartus Prime 软件第一次运行时需要配置 License,弹出如图 5-9 所示的配置请求对话框。选择最后一个选项"If you have a valid license file, specify the location of your license file",单击 OK 按钮,弹出如图 5-10 所示的对话框,在 License file 信息栏中选择 License.dat 文件所在的目录,作者计算机中的 License.dat 文件在"/home/intelFPGA/"目录下,因此该信息栏显示的信息为"/home/intelFPGA/License.dat"。设置好后,单击 OK 按钮,完成 License 配置。

图 5-9 License 配置请求对话框

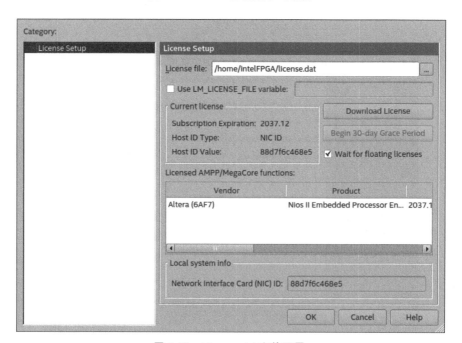

图 5-10 License.dat 文件配置

5.4.2　安装 SoCEDS

(1)～(7)重复 5.3.1 节安装步骤中的(1)～(7)。

(8) 赋予安装文件可执行权限，输入命令 chmod ＋ x SoCEDSSetup-18.1.0.625-linux.run。

(9) 执行安装，输入命令/SoCEDSSetup-18.1.0.625-linux.run。

在安装过程中会出现图 5-11(a)～图 5-11(h)的 8 个对话框，图 5-11(b)中需要选择"I accept the agreement"选项。

(a) 对话框1

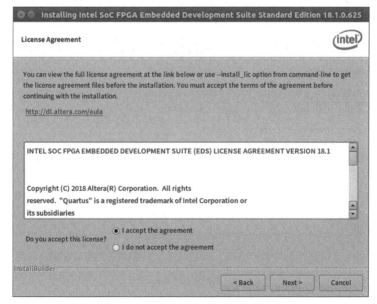

(b) 对话框2

图 5-11　SoCEDS 软件安装过程中出现的对话框

(c) 对话框3

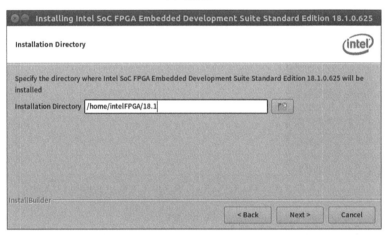

(d) 对话框4

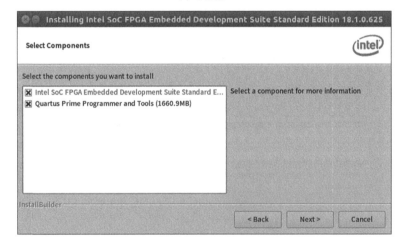

(e) 对话框5

图 5-11 （续）

(f) 对话框6

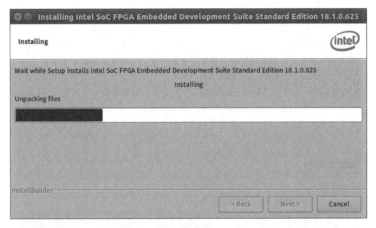

(g) 对话框7

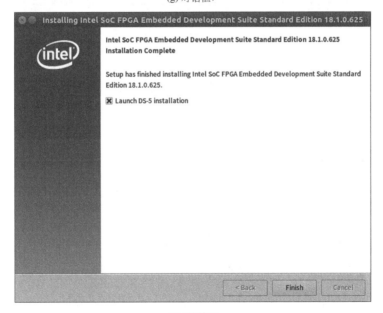

(h) 对话框8

图 5-11 （续）

注意：图 5-11(c)中的安装路径默认为"/root/intelFPGA/18.1"，因为"/root"目录采用图形界面访问不方便，因此需要将安装路径修改为"/home/intelFPGA/18.1"，如图 5-11(d)所示。同样，这样做也是为了与 Quartus Prime、Intel FPGA SDK for OpenCL 安装在同一目录下。

当单击图 5-11(h)中的 Finish 按钮后，弹出 ds-5 安装命令窗口，如图 5-12(a)所示。在弹出的命令窗口中连续按 Enter 键，直至出现如图 5-12(b)所示的界面。对给出的问题全部输入 yes，如图 5-12(c)所示。问题回答完成后，ds-5 安装命令窗口会自动关闭。

(a)

(b)

图 5-12　ds-5 安装命令窗口

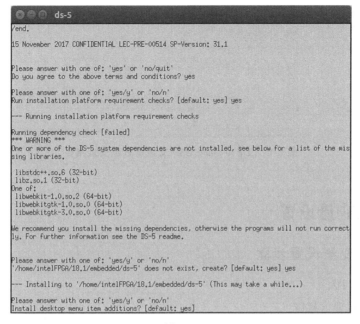

(c)

图 5-12 （续）

5.4.3 安装 DE10_nano BSP

1. DE10_nano BSP 下载

下载网址：http://www.terasic.com.cn/cgi-bin/page/archive.pl?Language=China&CategoryNo=203&No=1048&PartNo=4。

在 BSP for Intel FPGA SDK OpenCL 18.1 中选择下载 DE10-Nano OpenCL BSP（.tar.gz），如图 5-13 所示。

图 5-13　DE10_nano BSP 下载资源网站截图

2. DE10_nano BSP 安装

DE10-Nano_OpenCL_BSP_18.1.tar.gz 下载完成后，在"/home/intelFPGA/18.1/hld/board/"目录下新建文件夹并命名为 terasic。若无法创建新文件夹，则在 hld 目录下运行命令 chmod -R 777 board。将 DE10-Nano_OpenCL_BSP_18.1.tar.gz 解压到 terasic 目录下，生成名为 de10_nano 的文件夹，打开 de10_nano 文件夹，含有如图 5-14 所示的文件信息。

注意：图 5-14 中 DE10_nano 开发板的相关内容的路径为"/home/intelFPGA/18.1/hld/board/terasic/de10_nano"。

图 5-14 DE10_nano BSP 文件信息及安装路径

5.5 环境变量设置

5.5.1 环境变量设置步骤

（1）打开终端。
（2）获得管理员权限。
（3）运行终端命令 gedit /etc/profile。
（4）在打开的 profile 文档原有信息下面输入如下信息。

```
export ROOT=/home
export QUARTUS_ROOTDIR=$ROOT/intelFPGA/18.1/quartus
export INTELFPGAOCLSDKROOT=$ROOT/intelFPGA/18.1/hld
export PATH = $PATH:$QUARTUS_ROOTDIR/bin:$ROOT/intelFPGA/18.1/embedded/ds-5/
bin:$ROOT/intelFPGA/18.1/embedded/ds-5/sw/gcc/bin:$INTELFPGAOCLSDKROOT/
bin:$INTELFPGAOCLSDKROOT/arm32/bin
export LD_LIBRARY_PATH=$INTELFPGAOCLSDKROOT/arm32/lib
export AOCL_BOARD_PACKAGE_ROOT = $INTELFPGAOCLSDKROOT/board/terasic/de10
_nano
export QUARTUS_64BIT=1
export LM_LICENSE_FILE=$ROOT/intelFPGA/license.dat
```

（5）运行终端命令 source /etc/profile。

为了能够在 root 权限下运行 opencl SDK，有两种方法：一种是每次切换到 root 权限后运行命令 source /etc/profile；另一种是在"/root/.bashrc"文档中添加命令 source /etc/profile，这样可以避免每次切换到 root 权限后都要运行 source /etc/profile 命令。本书采用第二种方法。

（6）运行终端命令 gedit /root/.bashrc。
（7）在打开的 bashrc 文件原有信息的后面另起一行输入 source /etc/profile，如图 5-15 所示。
（8）运行终端命令 source /root/.bashrc。

```
# some more ls aliases
alias ll='ls -alF'
alias la='ls -A'
alias l='ls -CF'

# Alias definitions.
# You may want to put all your additions into a separate file like
# ~/.bash_aliases, instead of adding them here directly.
# See /usr/share/doc/bash-doc/examples in the bash-doc package.

if [ -f ~/.bash_aliases ]; then
    . ~/.bash_aliases
fi

# enable programmable completion features (you don't need to enable
# this, if it's already enabled in /etc/bash.bashrc and /etc/profile
# sources /etc/bash.bashrc).
#if [ -f /etc/bash_completion ] && ! shopt -oq posix; then
#    . /etc/bash_completion
#fi

export INTELFPGAOCLSDKROOT="/home/intelFPGA/18.1/hld"

export QSYS_ROOTDIR="/home/intelFPGA/18.1/quartus/sopc_builder/bin"

source /etc/profile
```

图 5-15 在"/root/.bashrc"文件中输入 source /etc/profile 命令

5.5.2 环境变量测试

（1）运行 aocl version 命令，显示 IntelFPGA SDK for OpenCL 的版本信息。图 5-16 为作者计算机上运行命令后的结果显示。

```
root@ubuntu602-System-Product-Name:/home/ubuntu602# aocl version
aocl 18.1.0.625 (Intel(R) FPGA SDK for OpenCL(TM), Version 18.1.0 Build 625 Stan
dard Edition, Copyright (C) 2018 Intel Corporation)
```

图 5-16 aocl version 命令运行结果

（2）运行 aoc -list-boards 命令，检测 DE10_nano 开发板信息。图 5-17 为作者计算机上运行命令后的结果显示。

```
root@ubuntu602-System-Product-Name:/home/ubuntu602# aoc -list-boards
Board list:
  de10_nano_sharedonly
     Board Package: /home/intelFPGA/18.1/hld/board/terasic/de10_nano
```

图 5-17 aoc -list-boards 命令运行结果

（3）运行 echo 命令，检测"/etc/profile"文件中的环境变量 AOCL_BOARD_PACKAGE_ROOT。图 5-18 为作者计算机上运行命令后的结果显示。

```
root@ubuntu602-System-Product-Name:/home/ubuntu602# echo $AOCL_BOARD_PACKAGE_ROOT
/home/intelFPGA/18.1/hld/board/terasic/de10_nano
```

图 5-18 echo $ AOCL_BOARD_PACKAGE_ROOT 命令运行结果

5.6 编译 OpenCL kernel

（1）打开终端。

（2）获得管理员权限。

（3）运行命令 cd /home/intelFPGA/18.1/hld/board/terasic/de10_nano/test/boardtest。

（4）运行命令 aoc boardtest.cl -o bin/boardtest.aocx -board=de10_nano_sharedonly -v -report。

在作者计算机上的运行命令及结果截图如图 5-19 所示。

```
root@ubuntu602-System-Product-Name:/home/ubuntu602# cd  /home/intelFPGA/18.1/hld/board/terasic/de10_na
no/test/boardtest
root@ubuntu602-System-Product-Name:/home/intelFPGA/18.1/hld/board/terasic/de10_nano/test/boardtest# ao
c boardtest.cl -o bin/boardtest.aocx -board=de10_nano_sharedonly -v -report
aoc: Environment checks are completed successfully.
aoc: Cached files in /var/tmp/aocl/root may be used to reduce compilation time
You are now compiling the full flow!!
aoc: Selected target board de10_nano_sharedonly
aoc: Running OpenCL parser....
/home/intelFPGA/18.1/hld/board/terasic/de10_nano/test/boardtest/boardtest.cl:78:20: warning: declaring
 kernel argument with no 'restrict' may lead to low kernel performance
    __global uint *dst,
                   ^
/home/intelFPGA/18.1/hld/board/terasic/de10_nano/test/boardtest/boardtest.cl:79:26: warning: declaring
 kernel argument with no 'restrict' may lead to low kernel performance
    __global const uint *index,
                         ^
2 warnings generated.
aoc: OpenCL parser completed successfully.
aoc: Optimizing and doing static analysis of code...
aoc: Linking with IP library ...
Checking if memory usage is larger than 100%

!=====================================================================
! The report below may be inaccurate. A more comprehensive
! resource usage report can be found at boardtest/reports/report.html
!=====================================================================

+--------------------------------------------------------------------+
; Estimated Resource Usage Summary                                   ;
+----------------------------------------+---------------------------+
; Resource                               + Usage                     ;
+----------------------------------------+---------------------------+
; Logic utilization                      ;    38%                    ;
; ALUTs                                  ;    22%                    ;
; Dedicated logic registers              ;    18%                    ;
; Memory blocks                          ;    32%                    ;
; DSP blocks                             ;    0%                     ;
+----------------------------------------+---------------------------;
aoc: First stage compilation completed successfully.
Compiling for FPGA. This process may take a long time, please be patient.
aoc: Hardware generation completed successfully.
```

图 5-19　boardtest.cl 内核编译结果

5.7 编译 host 程序

（1）打开终端。

（2）获得管理员权限。

（3）安装 lib32stdc++6,运行终端命令 apt-get install lib32stdc++6。

（4）安装 lib32z1,运行终端命令 apt-get install lib32z1(注意：结束字符为数字 1,不是英文字母 l)。

（5）切换终端工作目录到 /home/intelFPGA/18.1/embedded,运行终端命令 cd /home/intelFPGA/18.1/embedded。

（6）运行终端命令/embedded_command_shell.sh。

（7）切换终端工作目录到 /home/intelFPGA/18.1/hld/board/terasic/de10_nano/test/boardtest 运行终端命令 cd /home/intelFPGA/18.1/hld/board/terasic/de10_nano/test/boardtest。

（8）运行终端命令 make。

在作者电脑终端运行结果如图 5-20 所示。同时在/home/intelFPGA/18.1/hld/board/terasic/de10_nano/test/ boardtest 目录下会生成一个名为 boardtest_host 的文件。若之前该文件存在，注意该文件的修改时间，应该与当前时间很接近。

图 5-20　boardtest_host 程序编译结果

特别提醒：按照本教材安装步骤，host 代码中使用的头文件及对应的函数存放在 home/intelFPGA/18.1/hld/board/terasic/de10_nano/test/common/文件夹中。

为了能够正确编译 host 代码，需要在 home/intelFPGA/18.1/hld/board/terasic/de10_nano/test/文件夹下新建文件夹。若新建文件名为 design_1，则：

（1）在 design_1 文件下为 host 代码建立新的文件夹，并命名为 host。

（2）在 host 文件夹下新建文件夹，并命名为 src。

（3）在文件夹 src 下新建文本文档，并命名为 main.cpp。

（4）编写调试 main.cpp 文件。

（5）在 design_1 文件下为 kernel 代码建立新的文件夹，并命名为 device。

（6）在 device 文件夹下建立文本文档，并重命名为 * .cl(实际设计中需要用实际的名称替换 *)。

（7）复制开发板自带 Makefile 文件到 design_1 文件夹。

5.8　烧写 img 文件到 SD 卡（在 Windows 系统下完成）

本书使用的烧写软件为 Win32DiskImager(Win32 磁盘映像工具)，运行在 Windows 系统下。该软件可以从网站上自行下载。此外，还需要一个 USB 读卡器，需要将 SD 卡插入读卡器并通过 USB 接口连接到装有 Win32DiskImager 软件的计算机。

（1）打开 Win32DiskImager 软件，如图 5-21 所示。

图 5-21　Win32DiskImager 软件界面

（2）选择映像文件，并选择设备为读卡器对应的盘符。

如图 5-14 所示，DE10-Nano_OpenCL_BSP_18.1.tar.gz 解压后，在 de10_nano 文件夹下包含一个名为 de10_nano_opencl_18.1.img.zip 的文件，该文件即为需要写入 SD 卡的映像文件的压缩包。将该压缩包解压后得到的名为 de10_nano_opencl_18.1.img 文件即为映像文件。在图 5-21 中，"映像文件"一栏设置为该文件的路径信息即可。

作者计算机上 img 文件的所在路径及读卡器对应的盘符如图 5-22 所示。

图 5-22　映像文件及设备设置

设置映像文件后，单击"写入"按钮，会弹出如图 5-23 所示的"确认覆盖"对话框，单击 Yes 按钮，开始烧写 SD 卡。当烧写成功时，会弹出如图 5-24 所示的对话框，单击 OK 按钮，然后单击"退出"按钮，烧写 SD 卡完成。

图 5-23 "确认覆盖"对话框

图 5-24 SD 卡烧写成功

5.9 minicom 驱动安装与测试

5.9.1 minicom 驱动安装

（1）打开终端。

（2）获得管理员权限。

（3）运行命令 apt install minicom。

（4）运行命令 minicom -s，出现如图 5-25(a)所示界面。

（5）minicom 配置，用方向键选择 Serial port setup 选项，如图 5-25(b)所示，按 Enter 键，弹出如图 5-26(a)所示的界面。

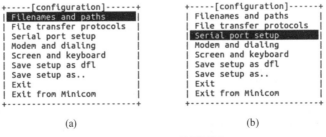

图 5-25 minicom 配置界面（1）

（6）从键盘输入 A（大小写均可），光标跳转到"A -Serial Device ：/dev/tty8"处，将 tty8 修改为 ttyUSB0，修改后按 Enter 键，光标跳回"Change which setting?"。从键盘输入 F（大小写均可），F-Hardware Flow Control：Yes 变为 No。设置好的界面如图 5-26（b）所示。

```
+--------------------------------------------------+
| A -    Serial Device      : /dev/tty8            |
| B -    Lockfile Location  : /var/lock            |
| C -    Callin Program     :                      |
| D -    Callout Program    :                      |
| E -    Bps/Par/Bits       : 115200 8N1           |
| F -    Hardware Flow Control : Yes               |
| G -    Software Flow Control : No                |
|                                                  |
|    Change which setting?                         |
+--------------------------------------------------+
```

(a)

```
+--------------------------------------------------+
| A -    Serial Device      : /dev/ttyUSB0         |
| B -    Lockfile Location  : /var/lock            |
| C -    Callin Program     :                      |
| D -    Callout Program    :                      |
| E -    Bps/Par/Bits       : 115200 8N1           |
| F -    Hardware Flow Control : No                |
| G -    Software Flow Control : No                |
|                                                  |
|    Change which setting?                         |
+--------------------------------------------------+
```

(b)

图 5-26   minicom 配置界面（2）

（7）在图 5-26（b）所示的界面下按 Enter 键，界面返回到如图 5-25（b）所示的界面。用方向键选择 Save setup as dfl 选项，如图 5-27（a）所示。按 Enter 键，出现 Configuration saved 信息提示。

（8）在图 5-27（a）所示的界面下用方向键选择 Exit from Minicom 选项，如图 5-27（b）所示。按 Enter 键退出 minicom 配置。

图 5-27   minicom 配置界面（3）

5.9.2   minicom 使用测试

本节将描述结合 DE10_nano 开发板验证 minicom 端口是否能够正常使用的方法。

（1）将烧写好的 SD 卡插入 DE10_nano 开发板的 SD 卡插槽中。

（2）将 miniUSB 线的 miniUSB 接口接到 DE10_nano 标有 UART 字样的插槽中（与

RJ-45 接口在同一侧)。

(3) 在计算机系统中打开终端。

(4) 获得管理员权限。

(5) 运行终端命令 minicom。

终端显示如图 5-28 所示的信息。

```
Welcome to minicom 2.7

OPTIONS: I18n
Compiled on Nov 15 2018, 20:18:47.
Port /dev/ttyUSB0, 09:47:16

Press CTRL-A Z for help on special keys
```

图 5-28 minicom 端口信息

(6) 开发板上电。如果 minicom 安装成功,则上电后终端窗口会显示 DE10_nano 开发板的启动信息。

(7) 登录开发板系统。当开发板启动后,需要登录才能进入系统,当出现 socfpga login: 命令行时,在命令行的后面输入 root,按 Enter 键。当出现 root@socfpga:~# 命令行时,说明成功进入开发板系统。

5.10　hello world kernel 运行测试

当进入开发板系统后,输入以下命令即可实现 hello world 程序的运行测试。

(1) 运行终端命令 source ./init_opencl.sh。

(2) 运行终端命令 ls。

(3) 运行终端命令 cd terasic/。

(4) 运行终端命令 ls。

(5) 运行终端命令 cd hello_world/。

(6) 运行终端命令 ls。

(7) 运行终端命令 ./host。

运行命令如图 5-29 所示,运行结果如图 5-30 所示,显示信息 Thread #2: Hello from Altera's OpenCL Compiler!。

```
socfpga login: root
root@socfpga:~# source ./init_opencl.sh
root@socfpga:~# ls
README              init_opencl.sh    terasic
altera              opencl_arm32_rte
root@socfpga:~# cd terasic/
root@socfpga:~/terasic# ls
boardtest    hello_world    vector_add
root@socfpga:~/terasic# cd hello_world/
root@socfpga:~/terasic/hello_world# ls
hello_world.aocx    host
root@socfpga:~/terasic/hello_world# ./host
```

图 5-29 hello world kernel 运行命令

```
Querying platform for info:
==========================
CL_PLATFORM_NAME                            = Intel(R) FPGA SDK for OpenCL(TM)
CL_PLATFORM_VENDOR                          = Intel(R) Corporation
CL_PLATFORM_VERSION                         = OpenCL 1.0 Intel(R) FPGA SDK for Ope1

Querying device for info:
========================
CL_DEVICE_NAME                              = de10_nano_sharedonly : Cyclone V SoCt
CL_DEVICE_VENDOR                            = Intel(R) Corporation
CL_DEVICE_VENDOR_ID                         = 4466
CL_DEVICE_VERSION                           = OpenCL 1.0 Intel(R) FPGA SDK for Ope1
CL_DRIVER_VERSION                           = 18.1
CL_DEVICE_ADDRESS_BITS                      = 64
CL_DEVICE_AVAILABLE                         = true
CL_DEVICE_ENDIAN_LITTLE                     = true
CL_DEVICE_GLOBAL_MEM_CACHE_SIZE             = 32768
CL_DEVICE_GLOBAL_MEM_CACHELINE_SIZE         = 0
CL_DEVICE_GLOBAL_MEM_SIZE                   = 536870912
CL_DEVICE_IMAGE_SUPPORT                     = true
CL_DEVICE_LOCAL_MEM_SIZE                    = 16384
CL_DEVICE_MAX_CLOCK_FREQUENCY               = 1000
CL_DEVICE_MAX_COMPUTE_UNITS                 = 1
CL_DEVICE_MAX_CONSTANT_ARGS                 = 8
CL_DEVICE_MAX_CONSTANT_BUFFER_SIZE          = 134217728
CL_DEVICE_MAX_WORK_ITEM_DIMENSIONS          = 3
CL_DEVICE_MEM_BASE_ADDR_ALIGN               = 8192
CL_DEVICE_MIN_DATA_TYPE_ALIGN_SIZE          = 1024
CL_DEVICE_PREFERRED_VECTOR_WIDTH_CHAR       = 4
CL_DEVICE_PREFERRED_VECTOR_WIDTH_SHORT      = 2
CL_DEVICE_PREFERRED_VECTOR_WIDTH_INT        = 1
CL_DEVICE_PREFERRED_VECTOR_WIDTH_LONG       = 1
CL_DEVICE_PREFERRED_VECTOR_WIDTH_FLOAT      = 1
CL_DEVICE_PREFERRED_VECTOR_WIDTH_DOUBLE     = 0
Command queue out of order?                 = false
Command queue profiling enabled?            = true
Using AOCX: hello_world.aocx
Reprogramming device [0] with handle 1

Kernel initialization is complete.
Launching the kernel...

Thread #2: Hello from Altera's OpenCL Compiler!

Kernel execution is complete.
```

图 5-30 hello world kernel 运行结果

5.11 DE10_nano 与 PC 数据交换

为了能够将在 PC 上编译好的 host 程序复制到开发板上,需要通过网口进行数据交换。需要的硬件设备如下:

(1) 两端具有 RJ-45 接口的连接线两条。
(2) 一台网络交换机。

此外,也可以使用一条两端具有 RJ-45 接口的网线直接连接 DE10_nano 开发板和 PC,但这种情况并非一定可行,与计算机配置有关,因此建议使用网络交换机实现 DE10_nano 开发板和 PC 的连接方式。

(1) 打开终端。
(2) 获得管理员权限。

（3）运行终端命令 ifconfig。

检测 PC 端的网卡信息，图 5-31 为作者 PC 端上检测到的网卡信息，网卡名称为 enp4s0。

```
root@ubuntu602:/home/ubuntu602# ifconfig
enp4s0    Link encap:以太网  硬件地址 88:d7:f6:c4:68:e5
          UP BROADCAST MULTICAST  MTU:1500  跃点数:1
          接收数据包:24841 错误:0 丢弃:0 过载:0 帧数:0
          发送数据包:2666 错误:0 丢弃:0 过载:0 载波:0
          碰撞:0 发送队列长度:1000
          接收字节:8370511 (8.3 MB)  发送字节:238009 (238.0 KB)

lo        Link encap:本地环回
          inet 地址:127.0.0.1  掩码:255.0.0.0
          inet6 地址: ::1/128 Scope:Host
          UP LOOPBACK RUNNING  MTU:65536  跃点数:1
          接收数据包:437 错误:0 丢弃:0 过载:0 帧数:0
          发送数据包:437 错误:0 丢弃:0 过载:0 载波:0
          碰撞:0 发送队列长度:1000
          接收字节:44038 (44.0 KB)  发送字节:44038 (44.0 KB)
```

图 5-31　网卡信息检测

（4）将 PC 连接到网络交换机的输出端口。

（5）运行终端命令 ifconfig enp4s0 192.168.199.103。设置网卡的 IP 地址（注：读者需要将此命令中的 enp4s0 替换为步骤（3）中检测到的网卡名称）。

（6）运行终端命令 ifconfig。

检测步骤（5）设置是否成功。若显示的"inet 地址"信息与步骤（5）命令中的地址信息一致，则说明 IP 地址设置成功。图 5-32 为作者 PC 端的运行结果。

```
root@ubuntu602:/home/ubuntu602# ifconfig
enp4s0    Link encap:以太网  硬件地址 88:d7:f6:c4:68:e5
          inet 地址:192.168.199.103  广播:192.168.199.255  掩码:255.255.255.0
          inet6 地址: fe80::5fc7:f567:8a4:8896/64 Scope:Link
          UP BROADCAST RUNNING MULTICAST  MTU:1500  跃点数:1
          接收数据包:24866 错误:0 丢弃:0 过载:0 帧数:0
          发送数据包:2777 错误:0 丢弃:0 过载:0 载波:0
          碰撞:0 发送队列长度:1000
          接收字节:8373601 (8.3 MB)  发送字节:252423 (252.4 KB)

lo        Link encap:本地环回
          inet 地址:127.0.0.1  掩码:255.0.0.0
          inet6 地址: ::1/128 Scope:Host
          UP LOOPBACK RUNNING  MTU:65536  跃点数:1
          接收数据包:509 错误:0 丢弃:0 过载:0 帧数:0
          发送数据包:509 错误:0 丢弃:0 过载:0 载波:0
          碰撞:0 发送队列长度:1000
          接收字节:48254 (48.2 KB)  发送字节:48254 (48.2 KB)
```

图 5-32　IP 地址设置

（7）将未上电的 DE10_nano 开发板连接到网络交换机的输出端口。

（8）使用 miniUSB 链接 DE10_nano 和 PC，miniUSB 端口连接开发板的 UART 接口。

（9）打开一个新的终端（与步骤（1）打开的终端不同），并获得管理员权限（从步骤（10）～步骤（18）的操作都在步骤（9）新打开的终端中执行）。

（10）在步骤（9）打开的终端中运行命令 minicom。

（11）给 DE10_nano 开发板上电，开发板系统启动。

（12）登录开发板系统，当出现 socfpga login：命令行时，在命令行的后面输入 root，按 Enter 键。

（13）运行终端命令 ifconfig。

检测 DE10_nano 开发板的网卡信息，图 5-33 为检测到的网卡信息，网卡名称为 eth0。

```
root@socfpga:~# ifconfig
eth0      Link encap:Ethernet  HWaddr 1a:23:02:bb:fe:2a
          UP BROADCAST MULTICAST  MTU:1500  Metric:1
          RX packets:0 errors:0 dropped:0 overruns:0 frame:0
          TX packets:4 errors:0 dropped:0 overruns:0 carrier:0
          collisions:0 txqueuelen:1000
          RX bytes:0 (0.0 B)  TX bytes:592 (592.0 B)
          Interrupt:152

lo        Link encap:Local Loopback
          inet addr:127.0.0.1  Mask:255.0.0.0
          inet6 addr: ::1/128 Scope:Host
          UP LOOPBACK RUNNING  MTU:65536  Metric:1
          RX packets:0 errors:0 dropped:0 overruns:0 frame:0
          TX packets:0 errors:0 dropped:0 overruns:0 carrier:0
          collisions:0 txqueuelen:0
          RX bytes:0 (0.0 B)  TX bytes:0 (0.0 B)
```

图 5-33　开发板网卡信息检测

（14）运行终端命令 ifconfig eth0 192.168.199.102。

给 DE10_nano 开发板的网卡分配 IP 地址。

（15）运行终端命令 ifconfig。

检测步骤(14)设置是否成功。若显示的"inet 地址"信息与步骤(14)命令中的地址信息一致，则说明 DE10_nano 开发板的网卡 IP 地址设置成功。图 5-34 为配置成功信息。

```
root@socfpga:~# ifconfig
eth0      Link encap:Ethernet  HWaddr 8e:2d:1b:6f:f3:83
          inet addr:192.168.199.102  Bcast:192.168.199.255  Mask:255.255.255.0
          UP BROADCAST RUNNING MULTICAST  MTU:1500  Metric:1
          RX packets:88 errors:0 dropped:0 overruns:0 frame:0
          TX packets:6 errors:0 dropped:0 overruns:0 carrier:0
          collisions:0 txqueuelen:1000
          RX bytes:11568 (11.2 KiB)  TX bytes:1248 (1.2 KiB)
          Interrupt:152

lo        Link encap:Local Loopback
          inet addr:127.0.0.1  Mask:255.0.0.0
          inet6 addr: ::1/128 Scope:Host
          UP LOOPBACK RUNNING  MTU:65536  Metric:1
          RX packets:0 errors:0 dropped:0 overruns:0 frame:0
          TX packets:0 errors:0 dropped:0 overruns:0 carrier:0
          collisions:0 txqueuelen:0
          RX bytes:0 (0.0 B)  TX bytes:0 (0.0 B)

root@socfpga:~#
```

图 5-34　开发板 IP 地址设置

（16）运行终端命令 ping 192.168.199.103。

验证 DE10_nano 与 PC 是否可通信，运行结果如图 5-35 所示。

（17）新建文件夹 my_boardtest，运行终端命令 mkdir my_boardtest。

（18）运行终端命令 ls。

```
root@socfpga:~# ping 192.168.199.103
PING 192.168.199.103 (192.168.199.103) 56(84) bytes of data.
64 bytes from 192.168.199.103: icmp_req=1 ttl=64 time=0.880 ms
64 bytes from 192.168.199.103: icmp_req=2 ttl=64 time=0.836 ms
64 bytes from 192.168.199.103: icmp_req=3 ttl=64 time=0.825 ms
64 bytes from 192.168.199.103: icmp_req=4 ttl=64 time=0.820 ms
64 bytes from 192.168.199.103: icmp_req=5 ttl=64 time=0.823 ms
64 bytes from 192.168.199.103: icmp_req=6 ttl=64 time=0.813 ms
64 bytes from 192.168.199.103: icmp_req=7 ttl=64 time=0.804 ms
64 bytes from 192.168.199.103: icmp_req=8 ttl=64 time=0.821 ms
64 bytes from 192.168.199.103: icmp_req=9 ttl=64 time=0.824 ms
64 bytes from 192.168.199.103: icmp_req=10 ttl=64 time=0.815 ms
64 bytes from 192.168.199.103: icmp_req=11 ttl=64 time=0.820 ms
64 bytes from 192.168.199.103: icmp_req=12 ttl=64 time=0.807 ms
^C
--- 192.168.199.103 ping statistics ---
12 packets transmitted, 12 received, 0% packet loss, time 10996ms
rtt min/avg/max/mdev = 0.804/0.824/0.880/0.018 ms
root@socfpga:~#
```

图 5-35　执行 ping 命令

查看文件夹是否成功创建，若显示的列表中存在 my_boardtest，则说明步骤(17)创建文件夹成功，如图 5-36 所示。

```
root@socfpga:~# mkdir my_boardtest
root@socfpga:~# ls
README            init_opencl.sh      opencl_arm32_rte
altera            my_boardtest        terasic
root@socfpga:~#
```

图 5-36　ls 信息显示

（19）运行终端命令 cd /home/intelFPGA/18.1/hld/board/terasic/de10_nano/test/boardtest/bin/（注意：步骤(19)～步骤(23)操作在步骤(1)打开的终端内执行）。

（20）运行终端命令 scp boardtest.aocx root@192.168.199.102:/home/root/my_boardtest，执行结果如图 5-37 所示。

```
root@ubuntu602:/home/ubuntu602# cd /home/intelFPGA/18.1/hld/board/terasic/de10_n
ano/test/boardtest/bin/
root@ubuntu602:/home/intelFPGA/18.1/hld/board/terasic/de10_nano/test/boardtest/b
in# ls
boardtest   boardtest.aoco   boardtest.aocx
root@ubuntu602:/home/intelFPGA/18.1/hld/board/terasic/de10_nano/test/boardtest/b
in# scp boardtest.aocx root@192.168.199.102:/home/root/my_boardtest
The authenticity of host '192.168.199.102 (192.168.199.102)' can't be establishe
d.
ECDSA key fingerprint is SHA256:0p0dsTqOv/FIHk4P+hqw1cLPnbjYYb3EzcXfmy5twNA.
Are you sure you want to continue connecting (yes/no)? yes
Warning: Permanently added '192.168.199.102' (ECDSA) to the list of known hosts.
boardtest.aocx                              100% 3248KB   3.2MB/s   00:01
root@ubuntu602:/home/intelFPGA/18.1/hld/board/terasic/de10_nano/test/boardtest/b
in#
```

图 5-37　aocx 文件复制

（21）运行终端命令 cd ..。
（22）运行终端命令 ls。
（23）运行终端命令 scp boardtest_host root@192.168.199.102:/home/root/my_boardtest。执行结果如图 5-38 所示。

（24）运行终端命令 cd my_boardtest（注意：步骤(24)～步骤(29)操作在步骤(9)打

```
root@ubuntu602:/home/intelFPGA/18.1/hld/board/terasic/de10_nano/test/boardtest/b
in# cd ..
root@ubuntu602:/home/intelFPGA/18.1/hld/board/terasic/de10_nano/test/boardtest#
ls
bin  boardtest.cl  boardtest_host  host  Makefile
root@ubuntu602:/home/intelFPGA/18.1/hld/board/terasic/de10_nano/test/boardtest#
scp boardtest_host root@192.168.199.102:/home/root/my_boardtest
boardtest_host                              100%   54KB  54.5KB/s   00:00
root@ubuntu602:/home/intelFPGA/18.1/hld/board/terasic/de10_nano/test/boardtest#
```

图 5-38 host 文件复制

开的终端内执行)。

(25) 运行终端命令 ls。

查看 boardtest.aocx 和 boardtest_host 是否成功复制到 my_boardtest 文件夹目录下。若复制成功,则显示如图 5-39 所示的结果。

(26) 运行终端命令 cd ..。

(27) 运行终端命令 source ./init_opencl.sh。

(28) 运行终端命令 cd my_boardtest。

(29) 运行终端命令 ./boardtest_host。

运行命令截图如图 5-40 所示。

```
root@socfpga:~# cd my_boardtest
root@socfpga:~/my_boardtest# ls
boardtest.aocx   boardtest_host
root@socfpga:~/my_boardtest#
```

图 5-39 查看文件复制结果

```
root@socfpga:~/my_boardtest# cd ..
root@socfpga:~# source ./init_opencl.sh
root@socfpga:~# cd my_boardtest
root@socfpga:~/my_boardtest# ./boardtest_host
```

图 5-40 执行终端命令

运行结果在显示一些测试信息后,最后将显示 BOARDTEST PASSED。

习题 5

5.1 搭建 OpenCL 开发平台除了必要的开发板外,还需要其他什么硬件设备的支持?

5.2 搭建 OpenCL 开发平台需要安装哪些软件工具?

5.3 基于 DE10_nano 开发板的 OpenCL 平台中,USB 串口和 RJ-45 网口的作用是什么?

5.4 环境变量测试主要采用哪三个命令?

5.5 kernel 程序编译的命令是什么?

5.6 在 DE10_nano 开发板上运行 kernel 代码,必须先初始化运行环境,这一步需要执行什么指令?

5.7 Ubuntu 系统下如何获取硬件的网卡信息? 如何为网卡设置 IP 地址?

5.8 使用什么命令可以实现开发板与 PC 之间的数据迁移?

第 6 章

单层神经网络算法模型的 FPGA 实现流程

本章以无隐形层的单层神经网络实现 MNIST 数字识别为例,介绍基于 OpenCL 的神经网络算法设计与 FPGA 实现流程。整个流程包括算法分析、算法的 TensorFlow 模型训练与测试、算法的 OpenCL 模型设计与编译、文件复制移植及算法的 FPGA 运行等细节。最后介绍通过 report.rtml 文件获取算法设计的相关信息。

6.1 基于 OpenCL 的神经网络算法设计与 FPGA 实现的基本流程

基于 OpenCL 的神经网络算法设计与 FPGA 实现的流程如图 6-1 所示。

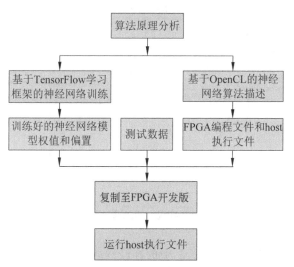

图 6-1 基于 OpenCL 的神经网络算法设计与 FPGA 实现的流程

(1) 对神经网络算法进行分析,明确其算法结构和原理。
(2) 在 TensorFlow 学习框架下设计该神经网络算法的 TensorFlow 模型,并对神经网络模型进行训练和测试,直至满足设计要求。当模型训练好后,保存模型的权值和偏置。

(3) 设计神经网络算法的 OpenCL 模型,即用 OpenCL 实现神经网络算法。借助于相关开发工具对设计进行编译,得到 FPGA 编程文件和 host 可执行文件。

(4) 将神经网络算法模型的相关文件移植到开发板。需要移植的文件包括训练好的模型的权值和偏置、FPGA 编程文件、host 可执行文件和测试样本集。

(5) 执行 host 可执行文件,运行神经网络算法模型。通过运行结果验证设计是否满足要求,并对设计进行优化和改进。

6.2 无隐形层的简易神经网络算法原理

图 6-2 为神经网络实现 MNIST 数据识别的示意图。MNIST 数据是 0~9 的数字图片,分为 28×28 个像素点,代表图像的尺寸。为了识别每个像素点,神经网络的权值个数为 784。此外,数据包含 0~9 共 10 个分类结果,根据全连接神经网络的结构原理,需要对每个分类结果分别提供 784 个权值,一般用权值个数衡量网络规模,因此该神经网络的规模为 784×10。

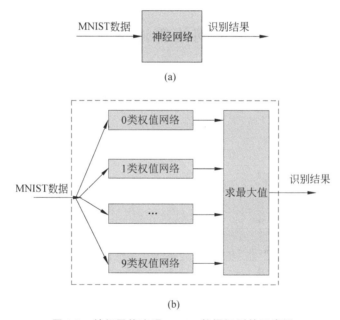

图 6-2 神经网络实现 MNIST 数据识别的示意图

图 6-2(b)的虚线框内对应图 6-2(a)中的神经网络部分,数字 0~9 分别对应一类权值网络。图 6-3 详细描述了每类权值网络的实现原理。$x[0] \sim x[783]$ 代表 784 个输入像素点,$w[0][k] \sim x[783][k]$ 代表第 k 类权值网络的权值,$b[k]$ 代表第 k 类权值网络的偏置,$y[k]$ 代表第 k 类权值网络的输出。

整个神经网络的权值个数为 784×10,在结构上为 784 行、10 列的数组,如图 6-4(a)

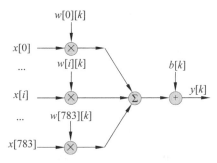

图 6-3 第 k 类权值网络（k 取值[0～9]）

所示。一般情况下，权值训练好后存储在存储器中。在读取数据时，为了方便算法实现，一般根据编号进行数据读取，实际的权值如图 6-4(b)所示。

$w[0][0]$	$w[0][1]$	$w[0][2]$	…	$w[0][9]$
$w[1][0]$	$w[1][1]$	$w[1][2]$	…	$w[1][9]$
…	…	…	…	…
$w[i][0]$	$w[i][1]$	$w[i][2]$	…	$w[i][9]$
…	…	…	…	…
$w[783][0]$	$w[783][1]$	$w[783][2]$	…	$w[783][9]$

(a)

$w[0]$	$w[1]$	$w[2]$	…	$w[9]$
$w[10]$	$w[11]$	$w[12]$	…	$w[19]$
…	…	…	…	…
$w[i\times10]$	$w[i\times10+1]$	$w[i\times10+2]$	…	$w[i\times10+9]$
…	…	…	…	…
$w[7830]$	$w[7831]$	$w[7832]$	…	$w[7839]$

(b)

图 6-4 权值的存储结构

根据神经网络结构及图 6-4 所示的权值存储结构，可以得到 10 类权值网络的算法描述如下。

```
for(int k=0;k< 10;k++)              //k 对应图 6-4 中的列数
{
    float y[k]={0.0};
    for (int i=0;i< 784;i++)        //i 对应图 6-4 中的行数
    {
        y[k]=y[k]+x[i] * w[k+i * 10];
    }
    y[k]=y[k]+b[k];
}
```

6.3 神经网络的 TensorFlow 实现及训练

（1）在 Ubuntu 系统的图形界面下，进入"/home/tensorflow/design/"文件夹，右击新建文件夹，并命名为 de10_nano_fpga。在文件夹 de10_nano_fpga 内再新建一个文件夹，并命名为 mnist_simple。

（2）在新建的 mnist_simple 文件夹内，右击新建空白文档，并重命名为 my_mnist_simple_fpga.py。

（3）打开(1)新建的 my_mnist_simple_fpga 文档，输入如下代码并保存。

```python
#coding: utf-8
import tensorflow as tf
import numpy as np
import sys
sys.path.append("/home/tensorflow/design/my_mnist/")
import pylab
import input_data
#载入数据集
Mnist=input_data.read_data_sets("/home/tensorflow/design/my_mnist/MNIST_data",one_hot=True)
#每个批次的大小
batch_size=100
#计算一共有多少个批次
n_batch=mnist.train.num_examples //batch_size
#定义两个placeholder
x=tf.placeholder(tf.float32,[None,784])
y=tf.placeholder(tf.float32,[None,10])
#创建一个简单的神经网络
W=tf.Variable(tf.zeros([784,10]))
b=tf.Variable(tf.zeros([10]))
prediction=tf.nn.softmax(tf.matmul(x,W)+b)
#使用交叉熵
loss=tf.reduce_mean(tf.nn.softmax_cross_entropy_with_logits(labels=y,logits=prediction))
#使用梯度下降法
train_step=tf.train.GradientDescentOptimizer(0.2).minimize(loss)
#初始化变量
init=tf.global_variables_initializer()
#结果存放在一个布尔型列表中
correct_prediction=tf.equal(tf.argmax(y,1),tf.argmax(prediction,1))
#argmax 返回一维张量中最大值所在的位置
#求准确率
accuracy=tf.reduce_mean(tf.cast(correct_prediction,tf.float32))
with tf.Session() as sess:
    sess.run(init)
    for epoch in range(101):
        for batch in range(n_batch):
            batch_xs,batch_ys= mnist.train.next_batch(batch_size)
            sess.run(train_step,feed_dict={x:batch_xs,y:batch_ys})
```

```
            train_acc=sess.run(accuracy,feed_dict=
                    {x:mnist.train.images,y:mnist.train.labels})
        print("Iter "+str(epoch)+",Training Accuracy "+str(train_acc))
            test_acc=sess.run(accuracy,feed_dict=
                    {x:mnist.test.images,y:mnist.test.labels})
 print("Testing Accuracy "+str(test_acc))
#显示第一张测试图片,用来验证 FPGA 运行的识别结果
img_in=mnist.test.images[0]
im=img_in.reshape(-1,28)
pylab.imshow(im)
pylab.show()
#将训练好的权值和偏置存储为 txt 文档
 w_sim_val, b_sim_val=sess.run([W,b])
 np.savetxt("w_sim.txt", w_sim_val.reshape(-1), fmt="%.31f", delimiter=",")
 np.savetxt("b_sim.txt", b_sim_val.reshape(-1), fmt="%.31f", delimiter=",")
 print("text write sucessful")
```

(4) 打开终端,获得管理员权限。

(5) 转换到 tensorflow 安装目录,运行终端命令 cd /home/tensorflow。

(6) 启动 tensorflow,运行终端命令 source bin/activate。

(7) 跳转到/home/tensorflow/design/de10_nano_fpga/mnist_simple/目录下,运行终端命令 cd /home/tensorflow/design/de10_nano_fpga/mnist_simple/。

(8) 运行 my_mnist_simple_fpga.py,运行终端命令 python my_mnist_simple_fpga.py。

运行结果如图 6-5 所示。图 6-5(b)显示第一张测试图片是 7,下文中在 FPGA 平台运行时使用第一张测试图片进行测试,若识别结果为 7,则可以验证 OpenCL 模型的正确性。

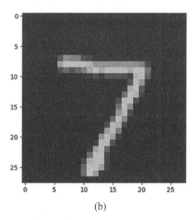

图 6-5 TensorFlow 模型运行结果

(9) 查看/home/tensorflow/design/de10_nano_fpga/mnist_simple/目录下同时生成的两个 txt 文件 b_sim.txt 和 w_sim.txt，它们分别保存了运行 my_mnist_simple_fpga.py 训练后的简易 MNIST 模型的偏置和权值。

注意：

① 运行该 py 文件的前提是已经运行第 3 章的 MNIST 例程。

```
sys.path.append("/home/tensorflow/design/my_mnist/")#该语句说明使用到了双引号中文件夹中的内容,该段代码中的input data文件和mnist数据存放在双引号中的文件夹中,若不存在,则请按照第3章的操作完成。
```

② b_sim.txt 和 w_sim.txt 需要写入 FPGA 开发板，这一操作将在 6.6 节介绍。

6.4 TensorFlow 框架下输入数据的转换

本书采用将图片转换为像素数据的方式进行处理，因此需要将 MNIST 测试数据集转换为 txt 文档格式。

（1）在 Ubuntu 系统的图形界面，在/home/tensorflow/design/de10_nano_fpga/mnist_simple/目录下新建文件夹，并命名为 mnist_txt。在文件夹 mnist_txt 内再新建两个文件夹，分别命名为 mnist_img_txt 和 mnist_lab_txt。

（2）在/home/tensorflow/design/de10_nano_fpga/mnist_simple/目录下新建空白文档，并命名为 input_file_gen.py，输入以下代码并保存。

```python
#coding: utf-8
import tensorflow as tf
import sys
sys.path.append("/home/tensorflow/design/my_mnist")
import input_data
#载入数据集
mnist=input_data.read_data_sets("../../MNIST_data",one_hot=True)
with tf.Session() as sess:
    import numpy as np
    for i in range(101):
        img_in_i=mnist.test.images[i]
        tag_i=np.argmax(mnist.test.labels[i])
        np.savetxt('./mnist_txt/mnist_img_txt/img_%d.txt'%i, img_in_i.reshape(-1), fmt="%.31f",delimiter=",")
        np.savetxt('./mnist_txt/mnist_lab_txt/img_lab_%d.txt'%i, tag_i.reshape(-1), fmt="%.31f",delimiter=",")
    print("text write sucessful")
```

（若 TensorFlow 已经在终端内打开，则忽略(3)~(5)）

（3）打开终端，获得管理员权限。

(4) 运行终端命令 cd /home/tensorflow,转换到 TensorFlow 安装目录。

(5) 运行终端命令 source bin/activate,启动 TensorFlow。

(6) 运行终端命令 python input_file_gen.py,执行 input_file_gen.py 文件。

(7) 查看/home/tensorflow/design/de10_nano_fpga/mnist_simple/mnist_txt/目录下的两个文件夹 mnist_img_txt 和 mnist_lab_txt 内分别生成的 101 个 txt 文件。其中,mnist_img_txt 文件夹中的 txt 文件名为 img_0~img100,mnist_lab_txt 文件夹中的 txt 文件名为 img_lab_0~img_lab_100。

6.5 神经网络算法的 OpenCL 实现

6.5.1 kernel 代码编写及编译

(1) 打开终端,获得管理员权限。

(2) 在/home/intelFPGA/18.1/hld/board/terasic/de10_nano/test/目录下新建文件夹 mnist_simple_one_image,在 mnist_simple_one_image 文件夹下新建文件夹 device,在 device 文件夹下新建空白文档,并将该文档重命名为 mnist_simple.cl,输入以下代码并保存。

```
__kernel void mnist_simple (__global const float *restrict dev_x, //28*28
                            __global const float *restrict dev_w, //784*10
                            __global const float *restrict dev_b,//10
                            __global float *restrict dev_y) //10
  {
    float rt[10]={0.0};
    for (unsigned thread_id = 0;thread_id<10;thread_id++)
    {
     for (int i=0;i<784;i++)
       {
          rt[thread_id]=rt[thread_id]+dev_x[i]*dev_w[thread_id+i*10];
       }
      dev_y[thread_id]=rt[thread_id]+dev_b[thread_id];
    }
  }
```

注意:由于代码编译依赖于/home/intelFPGA/18.1/hld/board/terasic/de10_nano/test/目录下的 common 文件夹,因此若读者自己选择新建文件夹的位置,则请将 common 文件夹复制到合适的目录下。

(3) 对 kernel 进行编译,运行终端命令 aoc device/mnist_simple.cl -o bin/mnist_simple.aocx -board=de10_nano_sharedonly -v -report。

终端运行结果如图 6-6 所示(编译 kernel 可能会消耗较长的时间,需要耐心等待)。

(4) 在 mnist_simple_one_image 文件下生成一个 bin 文件夹,文件夹中生成了一个名称为 mnist_simple 的文件夹和两个新文件 mnist_simple.aoco 与 mnist_simple.aocx,其中,mnist_simple.aocx 是 FPGA 的编程文件,需要复制到 FPGA 开发板上。mnist_simple 文件夹下包含了 kernel 编译的相关信息。

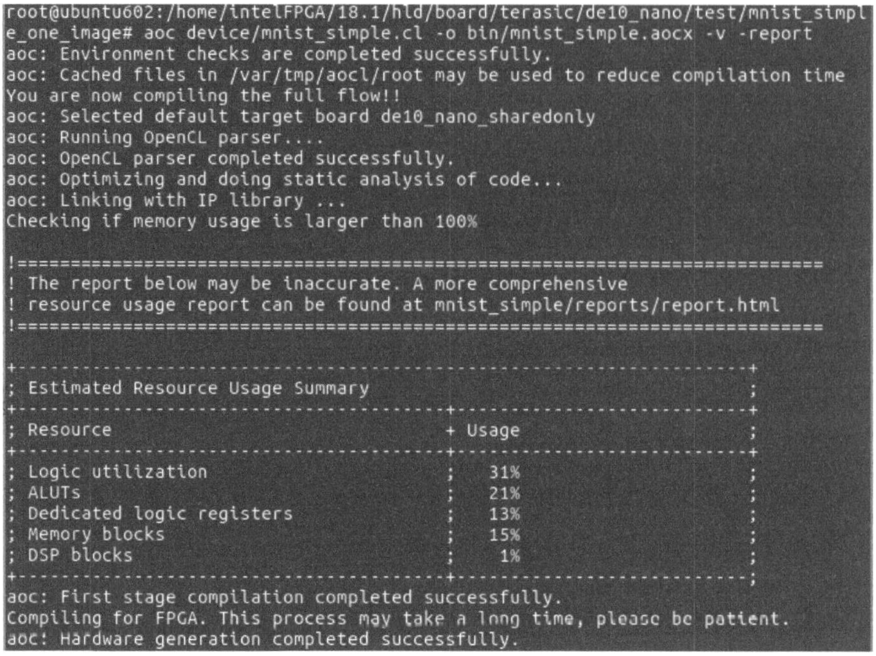

```
root@ubuntu602:/home/intelFPGA/18.1/hld/board/terasic/de10_nano/test/mnist_simpl
e_one_image# aoc device/mnist_simple.cl -o bin/mnist_simple.aocx -v -report
aoc: Environment checks are completed successfully.
aoc: Cached files in /var/tmp/aocl/root may be used to reduce compilation time
You are now compiling the full flow!!
aoc: Selected default target board de10_nano_sharedonly
aoc: Running OpenCL parser....
aoc: OpenCL parser completed successfully.
aoc: Optimizing and doing static analysis of code...
aoc: Linking with IP library ...
Checking if memory usage is larger than 100%

!===========================================================================
! The report below may be inaccurate. A more comprehensive
! resource usage report can be found at mnist_simple/reports/report.html
!===========================================================================

+--------------------------------------------------------------------+
; Estimated Resource Usage Summary                                   ;
+----------------------------------------+---------------------------+
; Resource                               + Usage                     ;
+----------------------------------------+---------------------------+
; Logic utilization                      ;   31%                     ;
; ALUTs                                  ;   21%                     ;
; Dedicated logic registers              ;   13%                     ;
; Memory blocks                          ;   15%                     ;
; DSP blocks                             ;    1%                     ;
+----------------------------------------+---------------------------;
aoc: First stage compilation completed successfully.
Compiling for FPGA. This process may take a long time, please be patient.
aoc: Hardware generation completed successfully.
```

图 6-6　kernel 编译结果

6.5.2　host 代码编写及编译

（1）在/home/intelFPGA/18.1/hld/board/terasic/de10_nano/test/mnist_simple_one_image 目录下新建文件夹 host。在 host 文件夹下新建文件夹 src，然后在 src 文件夹下新建空白文档，并命名为 main.cpp，输入以下代码。

```
#include <assert.h>
#include <stdio.h>
#include <stdlib.h>
#include <math.h>
#include <cstring>
#include "CL/opencl.h"
#include "AOCLUtils/aocl_utils.h"
#include <iostream>
#include <fstream>

#define image_size 784
#define x_size 28 * 28
#define w_size 784 * 10
#define b_size 10
#define y_size 10

using namespace aocl_utils;
```

```cpp
using namespace std;

//OpenCL runtime configuration
static cl_platform_id platform=NULL;
static cl_device_id device=NULL;
static cl_context context=NULL;
static cl_command_queue queue=NULL;
static cl_kernel mnist_kernel=NULL;
static cl_program program=NULL;

//Function prototypes
void ReadFloat(const char * filename,cl_float * data);
//
const char * kernel_name="mnist_simple";        //定义 kernel 名称
const char * source_file="mnist_simple.cl";     //定义 kernel 代码文件的名字
const char * aocx_file="mnist_simple";          //定义 FPGA 编程文件的名字
//设置测试图片及路径信息,选用测试集的第一张图片,图片中的数字信息为 7
const char * input_file_path="mnist_txt/mnist_img_txt/img_0.txt";
const char * input_label_path="mnist_txt/mnist_lab_txt/img_lab_0.txt";
cl_int status;
/////////////////////////////////////////////////////////////////////
int main() {
  //Get the OpenCL platform
clGetPlatformIDs(1, &platform, NULL);
   //  Obtain the available OpenCL devices.
clGetDeviceIDs(platform, CL_DEVICE_TYPE_ALL, 1,&device,NULL);
//Create the context.
  context=clCreateContext(NULL, 1, &device, NULL, NULL, &status);
//Create the command queue.
  queue= clCreateCommandQueue(context, device, CL_QUEUE_PROFILING_ENABLE,
&status);
//Create the program.
  std::string binary_file=getBoardBinaryFile(aocx_file, device);
  program=createProgramFromBinary(context, binary_file.c_str(), &device, 1);
//Build the program that was just created.
  status=clBuildProgram(program, 0, NULL, "", NULL, NULL);
//create the kernel
mnist_kernel=clCreateKernel(program, kernel_name, &status);
//allocate and initialize the input vectors
    cl_float * x, * w, * b;
    x=(cl_float *)alignedMalloc(sizeof(cl_float) * x_size);
                                      //为输入图像分配空间 28×28
    w=(cl_float *)alignedMalloc(sizeof(cl_float) * w_size);
                                      //为 w 分配空间 784×10
```

```c
    b=(cl_float *)alignedMalloc(sizeof(cl_float) * b_size);
                                                    //为bias分配空间10
//create the input buffer
    cl_mem dev_x, dev_w,dev_b,dev_y;
    dev_x=clCreateBuffer(context, CL_MEM_READ_WRITE, sizeof(cl_float) * x_size, NULL, &status);
    dev_w=clCreateBuffer(context, CL_MEM_READ_WRITE, sizeof(cl_float) * w_size, NULL, &status);
    dev_b=clCreateBuffer(context, CL_MEM_READ_WRITE, sizeof(cl_float) * b_size, NULL, &status);
    dev_y=clCreateBuffer(context, CL_MEM_READ_WRITE, sizeof(cl_float) * y_size, NULL, &status);
  //load data from text file
    ReadFloat(input_file_path,x);           //输入图像数据
    ReadFloat("w_sim.txt",w);               //输入权值数据
    ReadFloat("b_sim.txt",b);               //输入偏置数据
  //load label of input file
    cl_float * input_lab;
    input_lab=(cl_float *)alignedMalloc(sizeof(cl_float) * 1);
    ReadFloat(input_label_path,input_lab);  //输入图像标签
    printf("label_input is %d\n", (int)(*input_lab));
//Write buffer
    status=clEnqueueWriteBuffer(queue, dev_x, CL_TRUE, 0, sizeof(cl_float) * x_size, x, 0, NULL, NULL);
    status=clEnqueueWriteBuffer(queue, dev_w, CL_TRUE, 0, sizeof(cl_float) * w_size, w, 0, NULL, NULL);
    status=clEnqueueWriteBuffer(queue, dev_b, CL_TRUE, 0, sizeof(cl_float) * b_size, b, 0, NULL, NULL);

//set the arguments
  status=clSetKernelArg(mnist_kernel,0, sizeof(cl_mem), (void*)&dev_x);
  status=clSetKernelArg(mnist_kernel,1, sizeof(cl_mem), (void*)&dev_w);
  status=clSetKernelArg(mnist_kernel,2, sizeof(cl_mem), (void*)&dev_b);
  status=clSetKernelArg(mnist_kernel,3, sizeof(cl_mem), (void*)&dev_y);
//launch kernel
  cl_event event_kernel;
  static const size_t GSize[]={1};   //mnist_simple的global size
  static const size_t WSize[]={1};   //mnist_simple的local size
  status=clEnqueueNDRangeKernel(queue, mnist_kernel, 1, 0, GSize, WSize, 0, NULL, &event_kernel);
//read the output
    cl_float * y;
    y=(cl_float *)alignedMalloc(sizeof(cl_float) * y_size);
```

```
        status=clEnqueueReadBuffer(queue, dev_y, CL_TRUE, 0, sizeof(cl_float) * y
_size, y , 0, NULL, NULL);
  //display result
        for(int j=0;j<y_size;j++)
            {
                printf("j=%i,",j);
            printf("number=%.16f\n",y[j]);
              }
//display recognaize label
        float tmp=0;
          char lab;
          for(int j=0;j<10;j++)
            {
                if (y[j]>tmp)
                  {
                      tmp=y[j];
                      lab=j;
                  }

            }
  printf(" label_recognized=%i \n",lab);

/////////////////////////////
   clFlush(queue);
   clFinish(queue);
//device side
   clReleaseMemObject(dev_x);
   clReleaseMemObject(dev_w);
   clReleaseMemObject(dev_b);
   clReleaseMemObject(dev_y);
   clReleaseKernel(mnist_kernel);
   clReleaseProgram(program);
   clReleaseCommandQueue(queue);
   clReleaseContext(context);
//hose side
   free(x);
   free(w);
   free(b);
   free(y);
   free(input_lab);
/////////////////////////////
   return 0;
}
```

```
/////////////////////////////////////
void ReadFloat(const char * filename,cl_float * data)
{
    FILE * fp1;                        //定义文件流指针,用于打开读取的文件
    fp1=fopen(filename,"r+");          //读写方式打开文件
    int j=0;
    while(fscanf(fp1,"%f",&data[j++])!=-1);
                                       //逐行读取 fp1 所指文件中的内容到 data 中
    fclose(fp1);                       //关闭文件,有打开就要有关闭
}
/////////////////////////
void cleanup()
{
}
```

（2）从/home/intelFPGA/18.1/hld/board/terasic/de10_nano/test/hello_world 文件夹下复制文件 Makefile 到/home/intelFPGA/18.1/hld/board/terasic/de10_nano/test/mnist_simple_one_image 文件夹下。

注意：Makefile 是 BSP 自带的可完成 host 代码编译的批处理程序,请勿修改。

（3）打开终端,获得管理员权限。

（4）跳转到/home/intelFPGA/18.1/hld/board/terasic/de10_nano/test/mnist_simple_one_image/目录下。

（5）运行终端命令 make。

终端运行结果如图 6-7 所示。

```
root@ubuntu602:/home/intelFPGA/18.1/hld/board/terasic/de10_nano/test/mnist_simpl
e_one_image# make
arm-linux-gnueabihf-g++  -O2 -fPIC -I../common/inc \
               -I/home/intelFPGA/18.1/hld/host/include  host/src/main.cpp ../co
mmon/src/AOCLUtils/opencl.cpp ../common/src/AOCLUtils/options.cpp /home/intelFPG
A/18.1/hld/host/arm32/lib/libacl_emulator_kernel_rt.so /home/intelFPGA/18.1/hld/
host/arm32/lib/libOpenCL.so /home/intelFPGA/18.1/hld/host/arm32/lib/libalterahal
mmd.so /home/intelFPGA/18.1/hld/host/arm32/lib/libelf.so /home/intelFPGA/18.1/hl
d/host/arm32/lib/libc_accel_runtime.so /home/intelFPGA/18.1/hld/host/arm32/lib/l
ibalteracl.so /home/intelFPGA/18.1/hld/board/terasic/de10_nano/arm32/lib/libinte
l_soc32_mmd.so \
                \
                -lrt -lpthread \
                -o bin/host
```

图 6-7 make 命令运行结果

（6）查看/home/intelFPGA/18.1/hld/board/terasic/de10_nano/test/mnist_simple_one_image/bin/目录下是否有 host 文件生成,若有,则说明生成 host 代码编译成功。

6.6 数据移植复制到 FPGA 开发板

将 6.3 节生成的 b_sim.txt 和 w_sim.txt 文件,以及 6.4 节生成的 mnist_txt 文件夹中的所有 txt 文件复制到 DE10_nano 开发板的 SD 卡中。

(1)～(16)：参考5.9节的(1)～(16)。(10)～(17)的操作都在(9)新打开的终端中执行，即用户名为socfpga的终端，命令前导为root@socfpga的终端。

(17) 新建文件夹mnist_simple_one_image并查看结果。运行终端命令：

```
mkdir mnist_simple_one_image
ls
```

结果如图6-8所示。

```
root@socfpga:~# mkdir mnist_simple_one_image
root@socfpga:~# ls
README                  init_opencl.sh          opencl_arm32_rte
altera                  matrix_mult             pipe_test
arm_implementation      mnist_simple_one_image  terasic
fastica                 my_boardtest
fpga_implementaion      one_kernel_char
root@socfpga:~#
```

图6-8 新建文件夹mnist_simple_one_image

(18) 注意：(18)～(23)均在(1)打开的终端内执行，即用户名为Ubuntu系统用户名的终端，本书的用户名为ubuntu602，命令前导为root@ ubuntu602。

运行终端命令：

```
cd /home/intelFPGA/18.1/hld/board/terasic/de10_nano/test/mnist_simple_one_
image/bin/
```

(19) 运行终端命令：

```
scp mnist_simple.aocx root@192.168.199.102:/home/root/mnist_simple_one
_image
```

(20) 运行终端命令：

```
scp host root@192.168.199.102:/home/root/mnist_simple_one_image
```

(21) 运行终端命令：

```
cd /home/tensorflow/design/de10_nano_fpga/mnist_simple
```

(22) 运行终端命令：

```
scp -r mnist_txt root@192.168.199.102:/home/root/mnist_simple_one_image
```

(23) 运行终端命令：

```
scp b_sim.txt w_sim.txt root@192.168.199.102:/home/root/mnist_simple_one
_image
```

移植复制过程的终端运行结果如图 6-9 所示。

(a) FPGA编程文件和host可执行文件

(b) 测试样本集数据

(c) 测试样本集标签

图 6-9 数据移植复制结果

(24) 注意：(24)～(29)均在(9)打开的终端内执行，即用户名为 socfpga 的终端，命令前导为 root@socfpga。

运行终端命令：

```
cd mnist_simple_one_image/
```

(25) 查看 mnist_simple.aocx、host、b_sim.txt、w_sim.txt 和 mnist_txt 是否成功复制到 mnist_simple_one_image 文件夹目录下。运行终端命令：

```
ls
```

(26) 运行终端命令：

```
cd mnist_txt/
```

(27) 运行终端命令：

```
ls
```

(28) 运行终端命令：

```
cd mnist_img_txt/
```

(29) 运行终端命令：

```
ls
```

终端执行结果如图 6-10 所示。

```
root@socfpga:~# cd mnist_simple_one_image/
root@socfpga:~/mnist_simple_one_image# ls
b_sim.txt          mnist_simple.aocx  w_sim.txt
host               mnist_txt
root@socfpga:~/mnist_simple_one_image# cd mnist_txt/
root@socfpga:~/mnist_simple_one_image/mnist_txt# ls
mnist_img_txt  mnist_lab_txt
root@socfpga:~/mnist_simple_one_image/mnist_txt# cd mnist_img_txt/
root@socfpga:~/mnist_simple_one_image/mnist_txt/mnist_img_txt# ls
img_0.txt    img_23.txt  img_39.txt  img_54.txt  img_7.txt   img_85.txt
img_1.txt    img_24.txt  img_4.txt   img_55.txt  img_70.txt  img_86.txt
img_10.txt   img_25.txt  img_40.txt  img_56.txt  img_71.txt  img_87.txt
img_100.txt  img_26.txt  img_41.txt  img_57.txt  img_72.txt  img_88.txt
img_11.txt   img_27.txt  img_42.txt  img_58.txt  img_73.txt  img_89.txt
img_12.txt   img_28.txt  img_43.txt  img_59.txt  img_74.txt  img_9.txt
img_13.txt   img_29.txt  img_44.txt  img_6.txt   img_75.txt  img_90.txt
img_14.txt   img_3.txt   img_45.txt  img_60.txt  img_76.txt  img_91.txt
img_15.txt   img_30.txt  img_46.txt  img_61.txt  img_77.txt  img_92.txt
img_16.txt   img_31.txt  img_47.txt  img_62.txt  img_78.txt  img_93.txt
img_17.txt   img_32.txt  img_48.txt  img_63.txt  img_79.txt  img_94.txt
img_18.txt   img_33.txt  img_49.txt  img_64.txt  img_8.txt   img_95.txt
img_19.txt   img_34.txt  img_5.txt   img_65.txt  img_80.txt  img_96.txt
img_2.txt    img_35.txt  img_50.txt  img_66.txt  img_81.txt  img_97.txt
img_20.txt   img_36.txt  img_51.txt  img_67.txt  img_82.txt  img_98.txt
img_21.txt   img_37.txt  img_52.txt  img_68.txt  img_83.txt  img_99.txt
img_22.txt   img_38.txt  img_53.txt  img_69.txt  img_84.txt
```

图 6-10　查看 FPGA 开发板中的拷贝数据

6.7　FPGA 运行神经网络

(1) 确认当前终端的用户名为 socfpga，并切换至/home/root 路径，运行终端命令：

```
cd /home/root
```

(2) 初始化 OpenCL 运行环境，运行终端命令：

```
source ./init_opencl.sh
```

(3) 切换至 mnist_simple_one_image 目录下，运行终端命令：

```
cd mnist_simple_one_image
```

(4) 程序运行,运行终端命令:

```
./host
```

终端运行结果如图 6-11 所示。

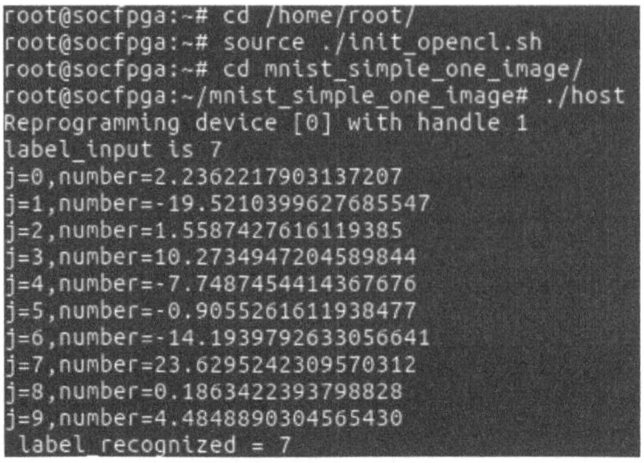

```
root@socfpga:~# cd /home/root/
root@socfpga:~# source ./init_opencl.sh
root@socfpga:~# cd mnist_simple_one_image/
root@socfpga:~/mnist_simple_one_image# ./host
Reprogramming device [0] with handle 1
label_input is 7
j=0,number=2.2362217903137207
j=1,number=-19.5210399627685547
j=2,number=1.5587427616119385
j=3,number=10.2734947204589844
j=4,number=-7.7487454414367676
j=5,number=-0.9055261611938477
j=6,number=-14.1939792633056641
j=7,number=23.6295242309570312
j=8,number=0.1863422393798828
j=9,number=4.4848890304565430
 label_recognized = 7
```

图 6-11　FPGA 运行结果

从终端运行结果可以看出,label_input 代表当前读取的图像的标签为 7,即该图像包含的信息为数字 7。number 代表当前图像经过神经网络处理后 0~9 十类数字的取值,对应算法中的 y(k),选取 10 组数据中最大的为识别结果,当 $j=7$ 时,结果最大。因此当前图像的识别结果为 label_recognized=7,与图 6-5(b)一致。

6.8　kernel report.html 文件查看

6.5.1 节中使用 aoc 命令对 kernel 代码进行编译时,在/home/intelFPGA/18.1/hld/board/terasic/de10_nano/ test/mnist_simple_one_image/bin 文件夹下生成了 mnist_simple 文件夹,该文件夹的 reports 文件夹中含有一个 report.html 文件,这个文件称为高层设计报告(High Level Design Report)。该文件包含 kernel 的分析数据,如 FPGA 资源和内存使用情况,以及循环结构和内核的流水信息。

本章以 6.5.1 节中的 kernel 代码生成的 report.html 为例,介绍使用该文件的方法。

6.8.1　高层设计报告布局

如图 6-12 所示,高层设计报告主要包含 4 个部分:报告菜单(report menu)、分析窗口(analysis pane)、源代码窗口(source code pane)和细节窗口(details pane)。

1. 报告菜单

如图 6-13 所示,从 Viewreports 的下拉菜单中可以选择查看 kernel 设计多个类别的分析信息,包括系统概要(Summary)、迭代分析(Loops analysis)、系统资源分析(Area analysis of system)、代码资源分析(Area analysis of source)、系统视图(System viewer)、内核内存视图(Kernel memory viewer)。

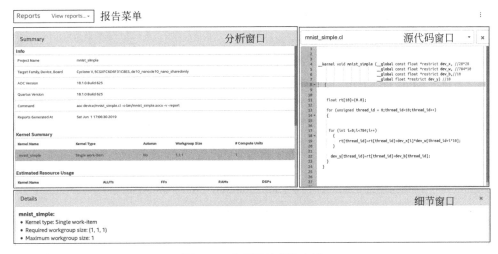

图 6-12　高层设计报告布局

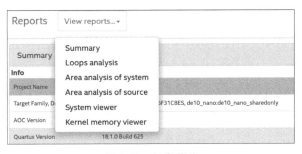

图 6-13　报告菜单

2. 分析窗口

分析窗口显示从 View reports 下拉菜单中选择的类别的详细信息。

3. 源代码窗口

源代码窗口显示所有源文件的代码。要想在内核中的不同源文件之间进行选择,可以单击该窗口顶部的下拉菜单。

4. 细节窗口

当选中出现在分析窗口中的某些信息时,该窗口将显示对应信息的细节说明。

6.8.2　系统概要

系统概要提供了对设计结果的快速概述,包括设计中每个内核的摘要及其使用的资源估计的摘要。如图 6-14 所示,系统概要分为 4 个部分:信息(Info)、内核摘要(Kernel Summary)、资源使用估计(Estimated Resource Usage)和编译警告(Compile Warnings)。

1. 信息

如图 6-14(a)所示,信息部分显示有关编译的一般信息,包括:

- 项目名称;

Info	
Project Name	mnist_simple
Target Family, Device, Board	Cyclone V, 5CSXFC6D6F31C8ES, de10_nano:de10_nano_sharedonly
AOC Version	18.1.0 Build 625
Quartus Version	18.1.0 Build 625
Command	aoc device/mnist_simple.cl -o bin/mnist_simple.aocx -v -report
Reports Generated At	Sat Jun 1 17:06:30 2019

(a) 信息

Kernel Summary				
Kernel Name	Kernel Type	Autorun	Workgroup Size	# Compute Units
mnist_simple	Single work-item	No	1,1,1	1

(b) 内核摘要

Estimated Resource Usage				
Kernel Name	ALUTs	FFs	RAMs	DSPs
mnist_simple	10709	15376	55	1
Global Interconnect	9588	10682	0	0
Board Interface	2160	1908	20	0
Total	22537 (21%)	27966 (13%)	75 (15%)	1 (1%)
Available	109572	219144	514	112

(c) 资源使用估计

Compile Warnings
None

(d) 编译警告

图 6-14 系统概要

- 目标 FPGA 系列、设备和主板；
- Intel Quartus® 主版本；
- AOC 版本；
- 用于编译设计的命令；
- 生成报告的日期和时间。

2. 内核摘要

如图 6-14(b)所示,内核摘要列出了设计中的每个内核,以及有关每个内核的一些信息,包括:

- 内核是 ndrange 还是 single work item；
- 是否使用 Autorun 属性；
- 内核所需的工作组大小；
- 计算单位的数量。

在列表中选择内核时,"详细信息"窗格将显示有关内核的其他信息,如图 6-15 所示。

图 6-15 内核摘要及细节窗口显示内容

3. 资源使用估计

如图 6-14(c)所示,资源使用估计部分显示设计中每个内核的资源使用估计摘要,以及所有通道(channel)、全局互连、常量缓存和板级接口的资源使用估计。

4. 编译警告

如图 6-14(d)所示,编译警告部分显示编译期间生成的一些编译器警告。

6.8.3 迭代分析

迭代分析(loop analysis)内容包含设计中所有循环及其展开状态。此循环迭代分析报告可帮助设计者检查离线编译器(Intel FPGA SDK for OpenCL)是否能够最大限度地提高内核的吞吐量。

设计者可以使用循环迭代分析报告确定在循环迭代代码中部署一个或多个 pragma 的位置,包含的 pragma 语句有♯pragma unroll、♯pragma loop_coalesce、♯pragma ii。

自动默认全部展开。kernel 程序的 mnist_simple.cl 中有两个迭代循环,分别对应代码的第 13 行和第 17 行的 for 循环。图 6-16(a)和(b)为对应第 13、17 行的 for 循环的信息,信息类别包括 pipelined、II、Bottleneck。细节窗口显示了循环迭代的详细信息。

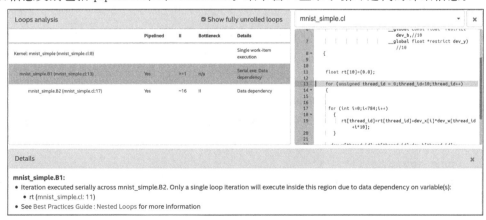

(a)

图 6-16 迭代分析及细节窗口显示的内容

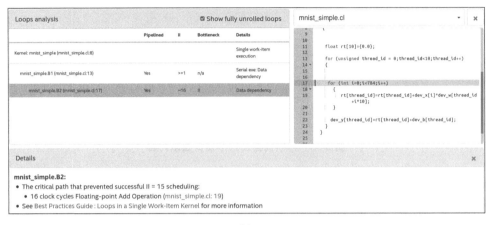

(b)

图 6-16 （续）

6.8.4 资源分析

资源分析（Area analysis of system & Area analysis of source）信息包含有关 OpenCL 系统资源使用的信息。

资源分析报告的用途如下：
- 提供整个 OpenCL 系统的详细资源划分，划分结果与源代码相关；
- 提供架构细节以深入了解生成的硬件，并提供可操作的建议以解决潜在的效率低下问题。

从报告菜单中的 View reports 的下拉菜单中选择 Area analysis by source 或者 Area analysis of system 项，可以按源代码或系统查看资源使用信息。资源使用数据是 Intel FPGA SDK for OpenCL 离线编译器生成的估计数据，这些估计值可能与最终的资源使用结果不同。

Area Analysis of System 报告显示 OpenCL 系统的资源分割，这是与在 FPGA 中实现的硬件最接近的。Area analysis by source 报告显示了源代码的每一行如何影响资源使用的估计值。

如图 6-17 所示，资源报告分为以下三个层次。
- 系统资源：给出了整个系统中所有内核、通道、互连和板级接口、逻辑等资源信息。图 6-17 中的系统资源由三部分组成：Global interconnect、Board interface 和内核 mnist_simpe。
- 内核资源：适用于特定内核，例如图 6-17 中的内核 mnist_simple，包括 Function overhead（功能开销），例如调度逻辑等。
- 基本块资源：适用于内核中的特定功能模块。基本块资源表示源代码中无分支的部分，例如循环体。图 6-18 为内核基本块展开后的资源使用信息。

第6章 单层神经网络算法模型的FPGA实现流程

Area analysis of system
(area utilization values are estimated)
Notation file:X > file:Y indicates a function call on line X was inlined using code on line Y.

	ALUTs	FFs	RAMs	DSPs	Details
▼ Kernel System 系统资源	22537 (21%)	27966 (13%)	75 (15%)	1 (1%)	
Global interconnect	9588	10682	0	0	• Global int… • See %L for…
Board interface	2160	1908	20	0	• Platform i…
▼ mnist_simple 内核资源	10780 (10%)	15376 (7%)	55 (11%)	1 (1%)	• Number of…
Function overhead	1574	1505	0	0	• Kernel dis…
Private Variable: - 'i' (mnist_simple.cl:17)	40	27	0	0	• Type: Regi… • 1 register…
Private Variable: - 'rr' (mnist_simple.cl:11)	400	2050	0	0	• Type: Regi… • 10 registe…
Private Variable: - 'thread_id' (mnist_simple.cl:13)	7	36	0	0	• Type: Regi… • 1 register…
▶ mnist_simple.B0 基本块资源	5 (0%)	2 (0%)	0 (0%)	0 (0%)	
▶ mnist_simple.B1	1052 (1%)	835 (0%)	0 (0%)	0 (0%)	
▶ mnist_simple.B2	5265 (5%)	5621 (3%)	26 (5%)	1 (1%)	
▶ mnist_simple.B3	2443 (2%)	5292 (2%)	29 (6%)	0 (0%)	
▶ mnist_simple.B4	3 (0%)	8 (0%)	0 (0%)	0 (0%)	

图 6-17 资源报告的三个层次

	ALUTs	FFs	RAMs	DSPs	
▼ mnist_simple.B2	5265 (5%)	5621 (3%)	26 (5%)	1 (1%)	
Cluster logic	828	271	0	0	• Logic requ…
▼ Computation	2168	3306	26	1	
▼ mnist_simple.cl:17	63	0	0	0	
11-bit Integer Add (x2)	8	0	0	0	
32-bit Integer Add (x3)	55	0	0	0	
▼ mnist_simple.cl:19	2101	3306	26	1	
32-bit Integer Add (x2)	44	0	0	0	
Floating-point Add	995	717	0	0	
Floating-point Multiply	274	240	0	1	
Load (x2)	788	2349	26	0	• Load uses … • Load uses …
▼ mnist_simple.cl:20	4	0	0	0	
11-bit Integer Add	4	0	0	0	
▶ Feedback	1174	1094	0	0	• Resources …
▶ State	1095	950	0	0	• Resources …

图 6-18 Area analysis by system 的基本块资源信息

比较图 6-17 和图 6-19 可以发现，Area analysis by system 与 Area analysis by source 在第三个层次上的信息显示存在区别。Area analysis by system 以基本块为单元进行资源信息显示，而 Area analysis by source 以源代码行号为单元进行资源信息显示。

图 6-20 为 Area analysis by source 第三层次包含的资源信息，该信息对应 mnist_simple.cl 程序的第 19 行。图 6-21 为在图 6-20 中选取第三层次 mnist_simple.cl:19 时在源代码窗口高亮对应的源代码。

图 6-19　Area analysis by source 报告信息

图 6-20　Area analysis by source 的基本块资源信息

图 6-21　在源代码窗口中高亮分析窗口对应的资源选项

6.8.5 系统视图

如图 6-22 所示,系统视图(system viewer)是 OpenCL 系统的交互式图形报告,它允许设计者查看诸如 load/store 的大小和类型、暂停和延迟等信息。系统视图显示 OpenCL 系统的抽象网络列表。通过系统视图查看 OpenCL 设计的图形表示可以验证内存复制,并识别任何可阻塞的加载和存储指令。

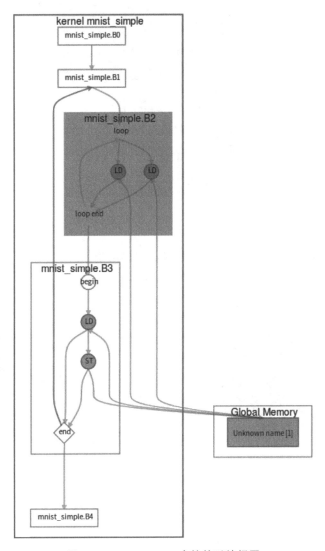

图 6-22 mnist_simple 内核的系统视图

检查与红色逻辑块关联的设计部分可以实现系统设计的性能改进。

例如,具有较大启动间隔(initiation interval,II)值的流水线循环的逻辑块可能以红色突出显示。启动间隔是指在流水线处理下一个循环迭代之前必须等待的硬件时钟周期数。最佳的循环展开后的 II 值为 1,因为每个时钟周期都需要处理一个循环迭代。大的

II值可能会影响设计吞吐量,如图 6-22 中的 mnist_simple.B2 为红色显示,说明这个逻辑块是影响系统性能的瓶颈所在。

将鼠标悬停在块中的任何节点上,可以在提示条和详细信息窗口中查看该节点的信息。如图 6-23 所示为当鼠标悬停在 mnist_simple.B2 逻辑块时显示的提示条信息。可以获知该逻辑块的延时为 210 个时钟周期,II 值为 16。

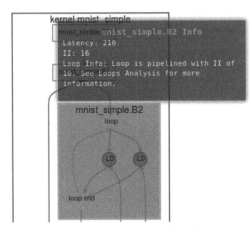

图 6-23　mnist_simple.B2 的整体信息

图 6-24 为当鼠标悬停在节点 LD 上时显示的提示条信息。

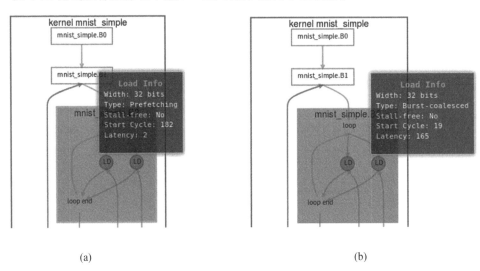

图 6-24　mnist_simple.B2 的 load 操作信息

图 6-24 中的 Width 代表数据的位数。Type 代表 LSU（Load-Store Unit,装载存储单元）的类型,编译器根据代码中的内存访问模式、可用内存类型以及内存访问是本地内存还是全局内存实例化 LSU 的类型。Stall-free 代表是否无阻塞,Start Cycle 代表开始执行的时钟周期,Latancy 代表该操作的延时。

图 6-24(a)中的 Type 是 Prefetching,图 6-24(b)中的 Type 是 Burst-coalesced。

Prefetching LSU 实例化一个 FIFO(有时称为 pipe),它根据以前的地址突发地从内存中读取大的块,以使 FIFO 充满有效数据并假定连续读取。支持非连续读取,但清空和重填 FIFO 会影响性能。

Burst-coalesced LSU 是编译器生成的默认 LSU 类型。该类型的 LSU 对请求进行缓冲,直到可能发生的最大突发。Burst-coalesced LSU 可以提供对全局内存的有效访问,但需要大量的 FPGA 资源。

此外,还有 Streaming LSU、Semi-streaming LSU、Local-Pipelined LSU、Global Infrequent、Constant-Pipelined LSU、Atomic-Pipelined LSU,读者可参阅相关用户手册了解各种 LSU 的特点。

6.9　log 文件查看 FPGA 资源使用估计信息

在 /home/intelFPGA/18.1/hld/board/terasic/de10_nano/test/mnist_simple_one_image/bin/mnist_simple 文件夹下含有名为 mnist_simple.log 的文件。该文件的第一行对应终端使用的 aoc 编译命令,用户主要通过该文件查看 FPGA 资源使用估计信息。

6.5.1 节的 kernel 代码的 FPGA 资源使用估计信息如图 6-25 所示。

```
+---------------------------------------------------+
; Estimated Resource Usage Summary                  ;
+---------------------------------+-----------------+
; Resource                        + Usage           ;
+---------------------------------+-----------------+
; Logic utilization               ;   31%           ;
; ALUTs                           ;   21%           ;
; Dedicated logic registers       ;   13%           ;
; Memory blocks                   ;   15%           ;
; DSP blocks                      ;    1%           ;
+---------------------------------+-----------------+
```

图 6-25　mnist_simple.log 文件中的资源估计信息

此外,在 mnist_simple 文件夹下还有一个名为 quartus_sh_compile.log 的文件,搜索关键词 logic cells 可以获取 quartus primer 软件的综合资源使用估计,如图 6-26 所示。

```
Implemented 13 input pins
Implemented 45 output pins
Implemented 57 bidirectional pins
Implemented 24468 logic cells
Implemented 2990 RAM segments
Implemented 1 PLLs
Implemented 1 delay-locked loops
Implemented 1 DSP elements
```

图 6-26　quartus_sh_compile.log 文件中的资源使用信息

习题 6

6.1　简述基于 OpenCL 的神经网络算法设计与 FPGA 实现的基本流程。
6.2　report.html 主要包含哪些信息?
6.3　report.html 中的系统概要包含哪四个部分?每一部分的作用是什么?

6.4 report.html 中的迭代分析的作用是什么？
6.5 report.html 中的资源分析的作用是什么？有哪两类资源分析？
6.6 report.html 中的资源分析报告的三个层次是什么？每个层次分别提供哪些信息？
6.7 report.html 中的系统视图的作用是什么？
6.8 如何查看 FPGA 资源使用信息？

第 7 章

单层神经网络算法的 kernel 程序实现方式分析比较

本章首先比较分析 kernel 编程方式、变量存储方式、数据类型对 FPGA 实现的资源使用、神经网络算法执行时间、FPGA 硬件运行效果的影响；然后比较分析 ARM 软件实现和 FPGA 硬件实现对 FPGA 实现的资源使用、神经网络算法执行时间、FPGA 硬件运行效果的影响。

7.1 批量读取输入数据的 OpenCL 程序

第 6 章介绍了 FPGA 运行神经网络算法读取一张测试图片进行测试的流程。本章的 host 代码可以读取多个输入图像文件，同时为对比不同 kernel 代码编写策略对 kernel 执行时间的影响而增加了时间测量函数。批量读取输入数据的 OpenCL 程序只需修改 host 程序代码，与 kernel 程序代码无关。

7.1.1 kernel 程序

同 6.5.1 节的 kernel 程序，继续采用 single work item 编程模式。

```
__kernel void mnist_simple (__global const float *restrict dev_x, //28*28
                            __global const float *restrict dev_w, //784*10
                            __global const float *restrict dev_b,//10
                            __global float *restrict dev_y) //10
{

  float rt[10]={0.0};
  for (unsigned thread_id = 0;thread_id<10;thread_id++)
  {

   for (int i=0;i<784;i++)
     {
      rt[thread_id]=rt[thread_id]+dev_x[i]*dev_w[thread_id+i*10];
     }
    dev_y[thread_id]=rt[thread_id]+dev_b[thread_id];
   }
}
```

7.1.2 host 程序

host 程序的代码如下：

```cpp
#include<assert.h>
#include<stdio.h>
#include<stdlib.h>
#include<math.h>
#include<cstring>
#include"CL/opencl.h"
#include"AOCLUtils/aocl_utils.h"
#include<iostream>
#include<fstream>

#define total_number_image 101
#define image_size 784
#define x_size 28*28
#define w_size 784*10
#define b_size 10
#define y_size 10

usingnamespace aocl_utils;
usingnamespace std;

//OpenCL runtime configuration
static cl_platform_id platform=NULL;
static cl_device_id device=NULL;
static cl_context context=NULL;
static cl_command_queue queue=NULL;
static cl_kernel mnist_kernel=NULL;
static cl_program program=NULL;

static float *x, *w, *b, *y=NULL;
static cl_mem dev_x,dev_w,dev_b,dev_y=NULL;

//Function prototypes
void ReadFloat(char * filename, cl_float * data);
double GetKernelExecutionTime(cl_command_queue cmd, cl_event event, char * eventname);

//
const char * kernel_name="mnist_simple";
const char * source_file="mnist_simple.cl";
```

```
const char * aocx_file="mnist_simple";
char input_file_name[5];
char suffix[5]=".txt";
int label_in[1];

cl_int status;
int img_rec_suc=0;
float correct_ratio=0;
float kernel_execution_time[total_number_image];

int main() {
//Get the OpenCL platform.
  clGetPlatformIDs(1, &platform, NULL);
//Query the available OpenCL devices.
clGetDeviceIDs(platform, CL_DEVICE_TYPE_ALL, 1,&device,NULL);
//Create the context.
  context=clCreateContext(NULL, 1, &device, NULL, NULL, &status);
//Create the command queue.
  queue=clCreateCommandQueue(context, device, CL_QUEUE_PROFILING_ENABLE,
&status);
//Create the program.
  std::string binary_file=getBoardBinaryFile(aocx_file, device);
  program=createProgramFromBinary(context, binary_file.c_str(), &device, 1);
//Build the program that was just created.
  status=clBuildProgram(program, 0, NULL, "", NULL, NULL);
/**********************************************************************/
double time1=getCurrentTimestamp();
/**********************************************************************/
//create the kernel
  mnist_kernel=clCreateKernel(program, kernel_name, &status);
//allocate and initialize the input vectors
    cl_float * x, * w, * b, * y;
    x=(cl_float *)alignedMalloc(sizeof(cl_float) * x_size);//
    w=(cl_float *)alignedMalloc(sizeof(cl_float) * w_size);//
    b=(cl_float *)alignedMalloc(sizeof(cl_float) * b_size);//
    y=(cl_float *)alignedMalloc(sizeof(cl_float) * y_size);//

//create the input buffer
    cl_mem dev_x,dev_w,dev_b,dev_y;

   dev_x=clCreateBuffer(context, CL_MEM_READ_WRITE, sizeof(cl_float) * x_
size, NULL, &status);
    dev_w=clCreateBuffer(context, CL_MEM_READ_WRITE, sizeof(cl_float) * w_
size, NULL, &status);
```

```
    dev_b=clCreateBuffer(context, CL_MEM_READ_WRITE, sizeof(cl_float) * b_
size, NULL, &status);
    dev_y=clCreateBuffer(context, CL_MEM_READ_WRITE, sizeof(cl_float) * y_
size, NULL, &status);

//load data from text file
    ReadFloat("w_sim.txt", w);
    ReadFloat("b_sim.txt", b);

//Write constant buffer
    status=clEnqueueWriteBuffer(queue, dev_w, CL_TRUE, 0, sizeof(cl_float) *
w_size, w, 0, NULL, NULL);
    status=clEnqueueWriteBuffer(queue, dev_b, CL_TRUE, 0, sizeof(cl_float) *
b_size, b, 0, NULL, NULL);

//read the input image file
for (int img_index=0;img_index<int(total_number_image);img_index++)
    {
        sprintf(input_file_name,"%d",img_index);

        char input_file_path[45]="./mnist_txt/mnist_img_txt/img_";
        char input_lab_path[45]="./mnist_txt/mnist_lab_txt/img_lab_";

    printf("##################################################\n");
    printf("    input_filename=\033[7mimg_%s.txt\033[0m\n",input_file_name);

    ReadFloat(strcat(strcat(input_file_path,input_file_name),suffix),x);//
    printf("**********************************************\n");
    printf("\033[7m \033[40;31m ******read image sucessful******************\
033[0m\n");
    printf("**********************************************\n");
////////////////////////////////////////////////////

    printf("**********************************************\n");
    printf("*   input_lab_filename=\033[7mimg_lab_%s.txt \033[0m\n",
input_file_name);
    printf("***********************************************\n");

//read the label of input image file

        FILE * fp1;
        fp1=fopen(strcat(strcat(input_lab_path,input_file_name),suffix),"r");
    //
```

```
    fscanf(fp1,"%i",label_in);
    fclose(fp1);
    printf("*********************************************\n");
    printf("*    \033[7m label of input image=%i    \033[0m\n",label_in[0]);
    printf("*********************************************\n");

//Write input image buffer
    status=clEnqueueWriteBuffer(queue, dev_x, CL_TRUE, 0, sizeof(cl_float) *
x_size, x, 0, NULL, NULL);

//set the arguments
   status=clSetKernelArg(mnist_kernel,0, sizeof(cl_mem), (void*)&dev_x);
   status=clSetKernelArg(mnist_kernel,1, sizeof(cl_mem), (void*)&dev_w);
   status=clSetKernelArg(mnist_kernel,2, sizeof(cl_mem), (void*)&dev_b);
   status=clSetKernelArg(mnist_kernel,3, sizeof(cl_mem), (void*)&dev_y);
//launch kernel
   cl_event event_kernel;
char dim=1;
static const size_t GSize[]={1};   //mnist_simple的global size
static const size_t WSize[]={1};   //mnist_simple的local size
   status=clEnqueueNDRangeKernel(queue, mnist_kernel, dim, 0, GSize, WSize,
0, NULL, &event_kernel);

/*************count runtime of kernel*************************/
double time_sum=0;
   time_sum=GetKernelExecutionTime(queue, event_kernel,"event_cluster");//
   kernel_execution_time[img_index]=time_sum/1000000;
/*****************************************************************/
//read the output
    status=clEnqueueReadBuffer(queue, dev_y, CL_TRUE, 0, sizeof(cl_float) *
y_size, y , 0, NULL, NULL);

//display result
       for(int j=0;j<y_size;j++)
         {
            printf("j=%i,",j);
            printf("number=%.16f\n",y[j]);
         }
//display recognaize label
       cl_float tmp=0;
       cl_float lab;

       for(int j=0;j<10;j++)
```

```
            {
                if(y[j]>tmp)
                {
                    tmp=y[j];
                    lab=j;
                }

            }
printf("*************************************************\n");
printf("*        \033[7m label_recognized=%i \033[0m\n",lab);
printf("*************************************************\n");
if (lab==label_in[0]) {
        img_rec_suc=img_rec_suc+1;
}


}

/**********************************************************************/

double time2=getCurrentTimestamp();
/**********************************************************************/
  correct_ratio=float(img_rec_suc)/int(total_number_image);
  printf("*************************************************\n");
  printf("*     img_input_num=%i\n",total_number_image);
  printf("*     img_rec_suc_num=\033[7m%i\033[0m        \n",img_rec_suc);
  printf("*     correct ratio=\033[7m%f\033[0m         \n",correct_ratio);
  printf("*************************************************\n");

/**********************************************************************/
//print runtime of kernel

printf("kernel_execution_time is:\n");
for (int i=0;i<total_number_image;i++)
printf("%f\n",kernel_execution_time[i]);

printf("total_time is:%f ms\n",(time2-time1) * 1e3);

///////////////////////////////
  clFlush(queue);
  clFinish(queue);
//device side
  clReleaseMemObject(dev_x);
  clReleaseMemObject(dev_w);
```

```
    clReleaseMemObject(dev_b);
    clReleaseMemObject(dev_y);
    clReleaseKernel(mnist_kernel);
    clReleaseProgram(program);
    clReleaseCommandQueue(queue);
    clReleaseContext(context);
//hose side
    free(x);
    free(w);
    free(b);
    free(y);

////////////////////////////
return 0;
}

//*************************************************************//
void ReadFloat(char * filename,cl_float * data)
{
    FILE * fp1;                //
    fp1=fopen(filename,"r+");  //
    int j=0;
    while(fscanf(fp1,"%f",&data[j++])!=-1);//
    fclose(fp1);               //
}

void cleanup()
{

}

double GetKernelExecutionTime(cl_command_queue cmd, cl_event event, char * eventname)
{
    cl_ulong start,end;

    clFinish(cmd);
    clGetEventProfilingInfo(event,CL_PROFILING_COMMAND_START,sizeof(cl_ulong),&start,NULL);
    clGetEventProfilingInfo(event,CL_PROFILING_COMMAND_END,sizeof(cl_ulong),&end,NULL);

    double runtime=(double)(end-start);

    return runtime;
}
```

7.1.3 执行结果

按照与 6.5~6.7 节相同的操作步骤完成相关操作。

注意：需要新建一个文件夹,防止与第一个实现混淆。由于 kernel 代码相同,因此可省略耗时的 aoc 命令,直接使用 aocx 文件即可,只需重新编译 host 文件,测试样本集数据和标签可以复制。

执行结果如图 7-1 所示。变量 img_input_num 表示输入的测试样本集数量；img_rec_suc_num 表示神经网络正确识别的测试样本集数量；correct ratio 表示识别成功率,为 img_rec_suc_num 和 img_input_num 的比值。从执行结果可以发现,在 FPGA 平台上运行的神经网络,当测试样本集数量为 101 时,可正确识别的样本数量为 97,识别成功率为 96.0396%。

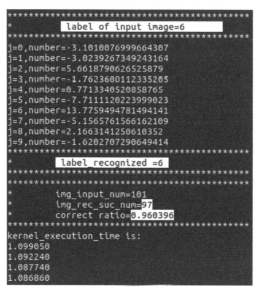

图 7-1 批量读取测试集运行结果

7.2 神经网络算法的不同 kernel 代码实现对比

7.2.1 single work item 和 NDRange(private)

本节分析 kernel 采用 single work item 编程模式和 NDRange 编程模式时,神经网络算法 FPGA 实现的几个方面的对比,包括 kernel 代码、FPGA 资源使用估计、kernel 执行时间以及 FPGA 运行结果。在此,private 指 kernel 代码中变量 rt 的存储模式,默认为 private 存储模式。

1. kernel 代码对比

图 7-2(a)为采用 single work item 编程模式实现的 kernel 程序,图 7-2(b)为采用

NDRange 编程模式实现的 kernel 程序。变量 rt 的存储类型默认为 private 类型。

```
__kernel void mnist_simple (__global const float *restrict dev_x, //28*28
                            __global const float *restrict dev_w, //784*10
                            __global const float *restrict dev_b, //10
                            __global float *restrict dev_y) //10
{
  float rt[10]={0.0};
  for (unsigned thread_id = 0;thread_id<10;thread_id++)
  {
    for (int i=0;i<784;i++)
      {
        rt[thread_id]=rt[thread_id]+dev_x[i]*dev_w[thread_id+i*10];
      }
     dev_y[thread_id]=rt[thread_id]+dev_b[thread_id];
  }
}
```

(a) single work item

```
__kernel void mnist_simple (__global const float *restrict dev_x, //28*28
                            __global const float *restrict dev_w, //784*10
                            __global const float *restrict dev_b, //10
                            __global float *restrict dev_y) //10
{
  float rt[10]={0.0};
  unsigned thread_id = get_global_id(0);
  for (int i=0;i<784;i++)
    {
      rt[thread_id]=rt[thread_id]+dev_x[i]*dev_w[thread_id+i*10];
    }
   dev_y[thread_id]=rt[thread_id]+dev_b[thread_id];
}
```

(b) NDRange

图 7-2 kernel 代码对比

2. FPGA 资源使用估计对比

在 mnist_simple.log 查看 FPGA 资源使用估计报告，图 7-3(a) 为 single work item 的 FPGA 资源使用估计信息，图 7-3(b) 为 NDRange 的 FPGA 资源使用估计信息。结果表明，本算法在 rt 使用 private 的情况下，NDRange 编程模式使用的 FPGA 资源更少。

```
+---------------------------------------------------+
; Estimated Resource Usage Summary                  ;
+---------------------------------+-----------------+
; Resource                        + Usage           ;
+---------------------------------+-----------------+
; Logic utilization               ;      31%        ;
; ALUTs                           ;      21%        ;
; Dedicated logic registers       ;      13%        ;
; Memory blocks                   ;      15%        ;
; DSP blocks                      ;       1%        ;
+---------------------------------+-----------------+
```

(a) single work item

图 7-3 FPGA 资源使用估计对比

```
+-----------------------------------------------------+
; Estimated Resource Usage Summary                    ;
+-----------------------------------------------------+
; Resource                          + Usage           ;
+-----------------------------------------------------+
; Logic utilization                 ;     24%         ;
; ALUTs                             ;     16%         ;
; Dedicated logic registers         ;      9%         ;
; Memory blocks                     ;     14%         ;
; DSP blocks                        ;      1%         ;
+-----------------------------------------------------+
```

(b) NDRange

图 7-3 （续）

3. kernel 运行时间对比

host 程序能够计算 kernel 每次运行的时间，并在 FPGA 终端运行时进行显示。根据显示结果，图 7-4(a) 给出了 single work item 模式 101 次 kernel 的运行时间，图 7-4(b) 给出了 NDRange 模式 101 次 kernel 的运行时间。结果对比表明，该情况下，NDRange 编程模式的 kernel 运行时间更短。

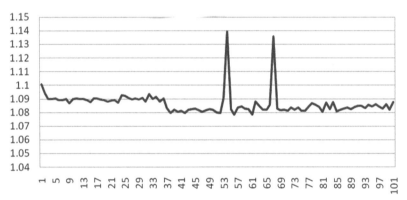

(a) single work item（单位：ms）

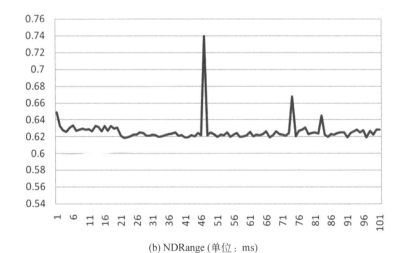

(b) NDRange（单位：ms）

图 7-4　kernel 运行时间对比

4. 运行结果对比

图 7-5 为终端显示的 FPGA 运行结果对比，图 7-5(a)为 single work item 编程模式的运行结果，图 7-5(b)为 NDRange 编程模式的运行结果。结果表明，二者的识别效果相同。

(a) single wrok item　　　　　　　　(b) NDRange

图 7-5　FPGA 运行结果对比

7.2.2　local 和 private(single work item)

本节分析比较 kernel 程序采用 single work item 模式下，rt 变量分别采用 local 和 private 存储模式的代码、资源使用估计、kernel 执行时间及 FPGA 运行结果的对比。

1. kernel 代码对比

图 7-6(a)为 single work item 编程模式下，rt 采用 private 存储类型实现的 kernel 程序，图 7-6(b)为 single work item 编程模式下，rt 采用 local 存储模式实现的 kernel 程序。

2. FPGA 资源使用估计对比

在 mnist_simple.log 中查看 FPGA 资源使用估计报告，图 7-7(a)为 private 存储模式的 FPGA 资源使用估计信息，图 7-7(b)为 local 存储模式的 FPGA 资源使用估计信息。结果表明，本算法在 single work item 编程模式下，rt 采用 local 存储模式时使用的 FPGA 资源更少。

3. kernel 运行时间对比

图 7-8(a)给出了 private 存储模式下 101 次 kernel 的运行时间，图 7-8(b)给出了 local 存储模式下 101 次 kernel 的运行时间。结果对比，该情况下，local 存储模式的 kernel 运行时间更少。

```
__kernel void mnist_simple (__global const float *restrict dev_x, //28*28
                            __global const float *restrict dev_w, //784*10
                            __global const float *restrict dev_b,//10
                            __global float *restrict dev_y) //10
{

  float rt[10]={0.0};

  for (unsigned thread_id = 0;thread_id<10;thread_id++)
  {

   for (int i=0;i<784;i++)
     {
        rt[thread_id]=rt[thread_id]+dev_x[i]*dev_w[thread_id+i*10];
     }

    dev_y[thread_id]=rt[thread_id]+dev_b[thread_id];
  }
}
```

(a) private

```
__kernel void mnist_simple (__global const float *restrict dev_x, //28*28
                            __global const float *restrict dev_w, //784*10
                            __global const float *restrict dev_b,//10
                            __global float *restrict dev_y) //10
{

  __local float rt[10];

  for (unsigned thread_id = 0;thread_id<10;thread_id++)
  {
   rt[thread_id]=0.0;
   for (int i=0;i<784;i++)
     {
        rt[thread_id]=rt[thread_id]+dev_x[i]*dev_w[thread_id+i*10];
     }

    dev_y[thread_id]=rt[thread_id]+dev_b[thread_id];
  }
}
```

(b) local

图 7-6 kernel 代码对比

```
+-------------------------------------------------------------+
; Estimated Resource Usage Summary                            ;
+--------------------------------------+----------------------+
; Resource                             + Usage                ;
+--------------------------------------+----------------------+
; Logic utilization                    ;   31%                ;
; ALUTs                                ;   21%                ;
; Dedicated logic registers            ;   13%                ;
; Memory blocks                        ;   15%                ;
; DSP blocks                           ;    1%                ;
+-------------------------------------------------------------+
```

(a) private

图 7-7 FPGA 资源使用估计对比

```
+---------------------------------------------+
; Estimated Resource Usage Summary            ;
+---------------------------------+-----------+
; Resource                        ; + Usage   ;
+---------------------------------+-----------+
; Logic utilization               ;   25%     ;
; ALUTs                           ;   17%     ;
; Dedicated logic registers       ;   10%     ;
; Memory blocks                   ;   15%     ;
; DSP blocks                      ;    1%     ;
+---------------------------------------------+
```

(b) local

图 7-7 （续）

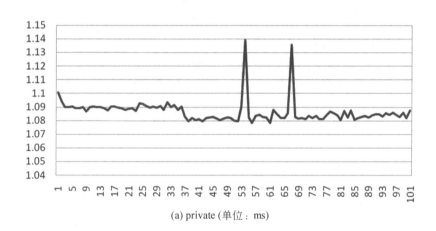

(a) private（单位：ms）

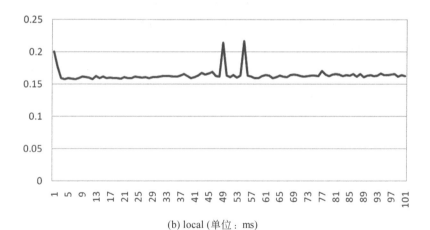

(b) local（单位：ms）

图 7-8　kernel 运行时间对比

4. 运行结果对比

图 7-9 为终端显示的 FPGA 运行结果对比，图 7-9(a)为 private 存储模式的运行结果，图 7-9(b)为 local 存储模式的运行结果。结果表明，二者的识别效果相同。

```
********************************************
*     input_lab_filename=img_lab_100.txt    *
********************************************
           label of input image=6
********************************************
j=0,number=-3.1010076999664307
j=1,number=-3.0239267349243164
j=2,number=5.6618790626525879
j=3,number=-1.7623608112335205
j=4,number=0.7713340520858765
j=5,number=-7.7111120223999023
j=6,number=13.7759494781494141
j=7,number=-5.1565761566162109
j=8,number=2.1663141250610352
j=9,number=-1.6202707290649414
********************************************
           label_recognized =6
********************************************
*         img_input_num=101                 *
*         img_rec_suc_num=97                *
*         correct ratio=0.960396            *
********************************************
kernel_execution_time is:
1.100610
1.093970
1.090020
```

(a) private

```
********************************************
*     input_lab_filename=img_lab_100.txt    *
********************************************
           label of input image=6
********************************************
j=0,number=-3.1010076999664307
j=1,number=-3.0239267349243164
j=2,number=5.6618790626525879
j=3,number=-1.7623608112335205
j=4,number=0.7713340520858765
j=5,number=-7.7111120223999023
j=6,number=13.7759494781494141
j=7,number=-5.1565761566162109
j=8,number=2.1663141250610352
j=9,number=-1.6202707290649414
********************************************
           label_recognized =6
********************************************
*         img_input_num=101                 *
*         img_rec_suc_num=97                *
*         correct ratio=0.960396            *
********************************************
kernel_execution_time is:
0.197850
0.178060
```

(b) local

图 7-9　FPGA 运行结果对比

7.2.3　local 和 private(NDRange)

7.2.2 节分析比较了 kernel 程序采用 single work item 编程模式下，rt 变量分别采用 local 和 private 存储模式的代码、资源使用估计、kernel 执行时间及 FPGA 运行结果的对比。本节分析比较 kernel 程序采用 NDRange 编程模式下，rt 变量分别采用 local 和 private 存储模式的代码、资源使用估计、kernel 执行时间及 FPGA 运行结果的对比。

1. kernel 代码对比

图 7-10(a)为 NDRange 编程模式下，rt 采用 private 存储模式实现的 kernel 程序，图 7-10(b)为 NDRange 编程模式下，rt 采用 local 存储模式实现的 kernel 程序。

```
__kernel void mnist_simple (__global const float *restrict dev_x, //28*28
                            __global const float *restrict dev_w, //784*10
                            __global const float *restrict dev_b,//10
                            __global float *restrict dev_y) //10
{

   float rt[10]={0.0};

   unsigned thread_id = get_global_id(0);

   for (int i=0;i<784;i++)
      {
         rt[thread_id]=rt[thread_id]+dev_x[i]*dev_w[thread_id+i*10];
      }

    dev_y[thread_id]=rt[thread_id]+dev_b[thread_id];
}
```

(a) private

图 7-10　kernel 代码对比

```
__kernel void mnist_simple (__global const float *restrict dev_x, //28*28
                            __global const float *restrict dev_w, //784*10
                            __global const float *restrict dev_b,//10
                            __global float *restrict dev_y) //10
{
   __local float rt[10];

   unsigned thread_id = get_global_id(0);

   rt[thread_id]=0.0;

   for (int i=0;i<784;i++)
      {
         rt[thread_id]=rt[thread_id]+dev_x[i]*dev_w[thread_id+i*10];
      }

    dev_y[thread_id]=rt[thread_id]+dev_b[thread_id];

}
```

(b) local

图 7-10 （续）

2. FPGA 资源使用估计对比

在 mnist_simple.log 中查看 FPGA 资源使用估计报告，图 7-11(a)为 private 存储模式的 FPGA 资源使用估计信息，图 7-11(b)为 local 存储模式的 FPGA 资源使用估计信息。结果表明，本算法在 NDRange 编程模式下，rt 采用 private 存储模式和 local 存储模式使用的 FPGA 资源相同，即在 NDRange 编程模式下，rt 采用 private 存储模式和 local 存储模式对 FPGA 资源的使用没有影响。

```
+-------------------------------------------------+
; Estimated Resource Usage Summary                ;
+---------------------------+---------------------+
; Resource                  + Usage               ;
+---------------------------+---------------------+
; Logic utilization         ;    24%              ;
; ALUTs                     ;    16%              ;
; Dedicated logic registers ;     9%              ;
; Memory blocks             ;    14%              ;
; DSP blocks                ;     1%              ;
+---------------------------+---------------------+
```

(a) private

```
+-------------------------------------------------+
; Estimated Resource Usage Summary                ;
+---------------------------+---------------------+
; Resource                  + Usage               ;
+---------------------------+---------------------+
; Logic utilization         ;    24%              ;
; ALUTs                     ;    16%              ;
; Dedicated logic registers ;     9%              ;
; Memory blocks             ;    14%              ;
; DSP blocks                ;     1%              ;
+---------------------------+---------------------+
```

(b) local

图 7-11 FPGA 资源使用估计对比

3. kernel 运行时间对比

图 7-12(a)给出了 private 存储模式下 101 次 kernel 的运行时间，图 7-12(b)给出了

local 存储模式下 101 次 kernel 的运行时间。结果对比,在 NDRange 编程模式下,local 存储模式的 kernel 运行时间基本一致。

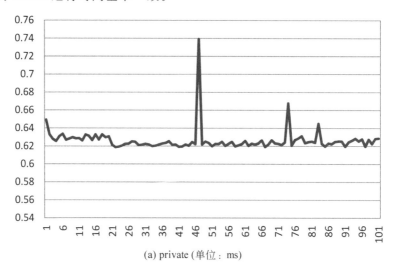

(a) private (单位:ms)

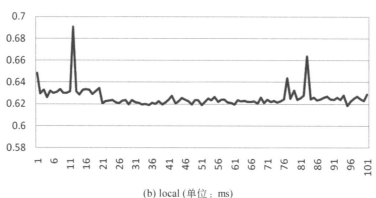

(b) local (单位:ms)

图 7-12　kernel 运行时间对比

4. 运行结果对比

图 7-13 为终端显示的 FPGA 运行结果对比,图 7-13(a)为 private 存储模式的运行结果,图 7-13(b)为 local 存储模式的运行结果。结果表明,二者的识别效果相同。

7.2.4　single work item 和 NDRange(local)

本节分析 kernel 采用 single work item 编程模式和 NDRange 编程方式时,rt 变量采用 local 存储模式时 kernel 代码、FPGA 资源使用估计、kernel 执行时间以及 FPGA 运行结果的对比。

1. kernel 代码对比

图 7-14(a)为 single work item 编程模式实现的 kernel 程序,图 7-14(b)为 NDRange 编程模式实现的 kernel 程序。

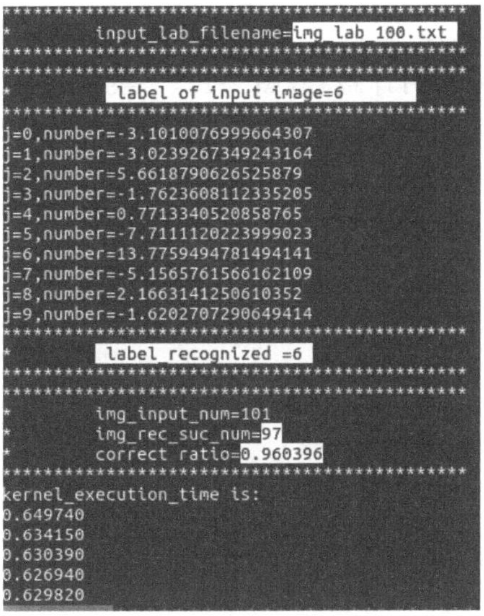

第7章 单层神经网络算法的kernel程序实现方式分析比较

```
***************************************
*          input_lab_filename=img_lab_100.txt
***************************************
*          label of input image=6
***************************************
j=0,number=-3.1010076999664307
j=1,number=-3.0239267349243164
j=2,number=5.6618790626525879
j=3,number=-1.7623608112335205
j=4,number=0.7713340520858765
j=5,number=-7.7111120223999023
j=6,number=13.7759494781494141
j=7,number=-5.1565761566162109
j=8,number=2.1663141250610352
j=9,number=-1.6202707290649414
***************************************
*          label_recognized =6
***************************************
           img_input_num=101
           img_rec_suc_num=97
           correct_ratio=0.960396
***************************************
kernel_execution_time is:
0.649740
0.634150
0.630390
0.626940
0.629820
```

(a) private

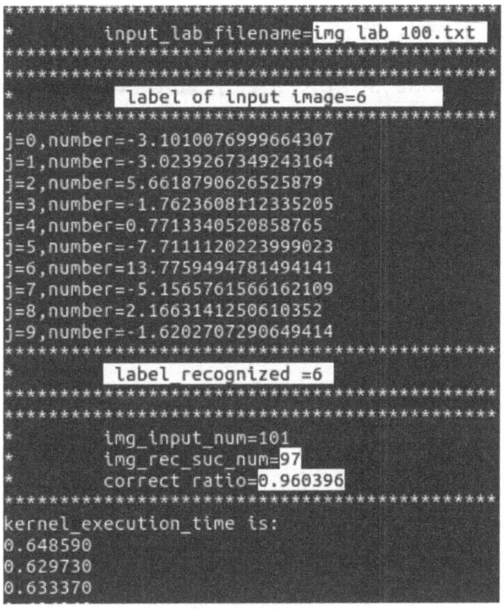

```
***************************************
*          input_lab_filename=img_lab_100.txt
***************************************
*          label of input image=6
***************************************
j=0,number=-3.1010076999664307
j=1,number=-3.0239267349243164
j=2,number=5.6618790626525879
j=3,number=-1.7623608112335205
j=4,number=0.7713340520858765
j=5,number=-7.7111120223999023
j=6,number=13.7759494781494141
j=7,number=-5.1565761566162109
j=8,number=2.1663141250610352
j=9,number=-1.6202707290649414
***************************************
*          label_recognized =6
***************************************
           img_input_num=101
           img_rec_suc_num=97
           correct_ratio=0.960396
***************************************
kernel_execution_time is:
0.648590
0.629730
0.633370
```

(b) local

图 7-13　FPGA 运行结果对比

```
__kernel void mnist_simple (__global const float *restrict dev_x, //28*28
                            __global const float *restrict dev_w, //784*10
                            __global const float *restrict dev_b,//10
                            __global float *restrict dev_y) //10
  {

    __local float rt[10];

    for (unsigned thread_id = 0;thread_id<10;thread_id++)
    {
      rt[thread_id]=0.0;

      for (int i=0;i<784;i++)
        {
          rt[thread_id]=rt[thread_id]+dev_x[i]*dev_w[thread_id+i*10];
        }

      dev_y[thread_id]=rt[thread_id]+dev_b[thread_id];
    }
  }
```

(a) single work item

```
__kernel void mnist_simple (__global const float *restrict dev_x, //28*28
                            __global const float *restrict dev_w, //784*10
                            __global const float *restrict dev_b,//10
                            __global float *restrict dev_y) //10
  {
    __local float rt[10];

    unsigned thread_id = get_global_id(0);

    rt[thread_id]=0.0;

    for (int i=0;i<784;i++)
      {
        rt[thread_id]=rt[thread_id]+dev_x[i]*dev_w[thread_id+i*10];
      }

    dev_y[thread_id]=rt[thread_id]+dev_b[thread_id];
  }
```

(b) NDRange

图 7-14　kernel 代码对比

2. FPGA 资源使用估计对比

在 mnist_simple.log 中查看 FPGA 资源使用估计报告，图 7-15(a)为 single work item 的 FPGA 资源使用估计信息，图 7-15(b)为 NDRange 的 FPGA 资源使用估计信息。结果表明，本算法在 rt 使用 local 的情况下，NDRange 编程模式使用的 FPGA 资源更少，但差别不大，前四项仅相差 1%。

```
+-------------------------------------------------+
; Estimated Resource Usage Summary                ;
+-------------------------------------------------+
; Resource                        + Usage         ;
;                                 ;               ;
; Logic utilization               ;    25%        ;
; ALUTs                           ;    17%        ;
; Dedicated logic registers       ;    10%        ;
; Memory blocks                   ;    15%        ;
; DSP blocks                      ;     1%        ;
+-------------------------------------------------+
```

(a) single work item

```
+-------------------------------------------------+
; Estimated Resource Usage Summary                ;
+-------------------------------------------------+
; Resource                        + Usage         ;
;                                 ;               ;
; Logic utilization               ;    24%        ;
; ALUTs                           ;    16%        ;
; Dedicated logic registers       ;     9%        ;
; Memory blocks                   ;    14%        ;
; DSP blocks                      ;     1%        ;
+-------------------------------------------------+
```

(b) NDRange

图 7-15　FPGA 资源使用估计对比

3. kernel 运行时间对比

图 7-16(a)给出了 single work item 模式下 101 次 kernel 的运行时间，图 7-16(b)给出了 NDRange 模式下 101 次 kernel 的运行时间。结果对比，该情况下，single work item 编程模式的 kernel 运行时间更少。

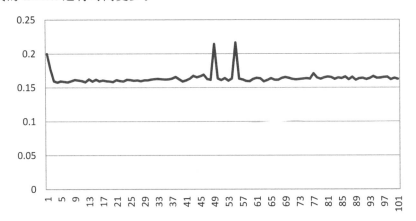

(a) single work item(单位：ms)

图 7-16　kernel 运行时间对比

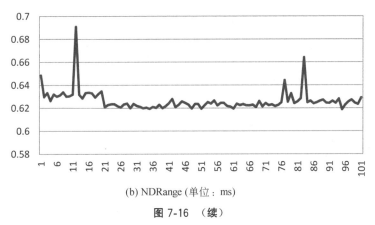

(b) NDRange（单位：ms）

图 7-16 （续）

4. 运行结果对比

图 7-17 为终端显示的 FPGA 运行结果对比，图 7-17(a) 为 single work item 编程模式的运行结果，图 7-17(b) 为 NDRange 编程模式的运行结果。结果表明，二者的识别效果相同。

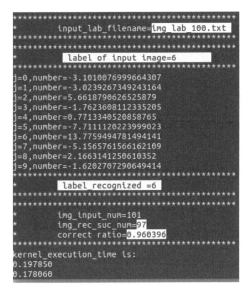

(a) single work item

(b) NDRange

图 7-17 FPGA 运行结果对比

小结 1：通过 7.2.1~7.2.4 节的对比分析可以发现，在采用 single work item 编程模式、rt 变量采用 local 存储模式情况下，kernel 的执行时间最短，使用的 FPGA 资源仅比 NDRange 编程模式多 1%。可以推断，针对本书给出的单层神经网络，single work item-local 是最佳实现方式。

7.2.5　float 和 char(single work item-local)

本节在 kernel 采用 single work item 编程模式，rt 变量采用 local 存储模式的情况下，分析比较 float 数据类型与 char 类型对 FPGA 实现的影响。

1. kernel 代码对比

图 7-18(a)为采用 float 数据类型实现的 kernel 程序，图 7-18(b)为采用 char 数据类型式实现的 kernel 程序。float 数据类型位数为 32 位，char 数据类型位数为 8 位。

```
__kernel void mnist_simple (__global const float *restrict dev_x, //28*28
                            __global const float *restrict dev_w, //784*10
                            __global const float *restrict dev_b,//10
                            __global float *restrict dev_y) //10
 {

  __local float rt[10];
  for (unsigned thread_id = 0;thread_id<10;thread_id++)
  {
   rt[thread_id]=0.0;
   for (int i=0;i<784;i++)
     {
        rt[thread_id]=rt[thread_id]+dev_x[i]*dev_w[thread_id+i*10];
     }
    dev_y[thread_id]=rt[thread_id]+dev_b[thread_id];
  }
 }
```

(a) float

```
__kernel void mnist_simple (__global const char *restrict dev_x, //28*28
                            __global const char *restrict dev_w, //784*10
                            __global const char *restrict dev_b,//10
                            __global char *restrict dev_y) //10
 {

    __local char rt[10];
    for (unsigned thread_id=0;thread_id<10;thread_id++)
       {
         rt[thread_id]=0;
         for (int i=0;i<784;i++)
            {
               rt[thread_id]=rt[thread_id]+dev_x[i]*dev_w[thread_id+i*10];
            }
          dev_y[thread_id]= rt[thread_id]+dev_b[thread_id];
       }
 }
```

(b) char

图 7-18　kernel 代码对比

2. FPGA 资源使用估计对比

在 mnist_simple.log 中查看 FPGA 资源使用估计报告，图 7-19(a)为 floats 数据类型的 FPGA 资源使用估计信息，图 7-19(b)为 char 数据类型的 FPGA 资源使用估计信息。结果表明，char 数据类型使用的 FPGA 资源更少。

```
+----------------------------------------------------+
; Estimated Resource Usage Summary                   ;
+----------------------------------------+-----------+
; Resource                               ; + Usage   ;
+----------------------------------------+-----------+
; Logic utilization                      ;   25%     ;
; ALUTs                                  ;   17%     ;
; Dedicated logic registers              ;   10%     ;
; Memory blocks                          ;   15%     ;
; DSP blocks                             ;   1%      ;
+----------------------------------------+-----------+
```

(a) float

```
+----------------------------------------------------+
; Estimated Resource Usage Summary                   ;
+----------------------------------------+-----------+
; Resource                               ; + Usage   ;
+----------------------------------------+-----------+
; Logic utilization                      ;   22%     ;
; ALUTs                                  ;   15%     ;
; Dedicated logic registers              ;   9%      ;
; Memory blocks                          ;   15%     ;
; DSP blocks                             ;   1%      ;
+----------------------------------------+-----------+
```

(b) char

图 7-19　FPGA 资源使用估计对比

3. kernel 运行时间对比

图 7-20(a)给出了 float 数据类型的 101 次 kernel 的运行时间，图 7-20(b)给出了 char 数据类型的 101 次 kernel 的运行时间。结果对比，该情况下，char 数据类型的 kernel 运行时间更短。

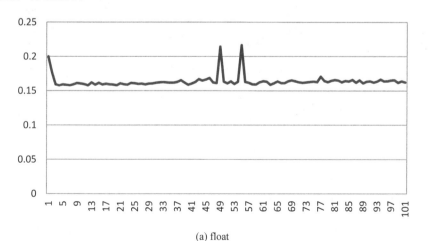

(a) float

图 7-20　kernel 运行时间对比

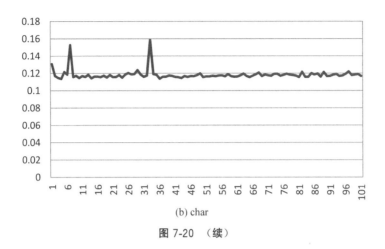

(b) char

图 7-20 （续）

4. 运行结果对比

图 7-21 为终端显示的 FPGA 运行结果对比，图 7-21(a) 为 float 数据类型的运行结果，图 7-21(b) 为 char 数据类型的运行结果。结果表明，char 数据类型略差于 float 数据类型。

(a) float   (b) char

图 7-21 FPGA 运行结果对比

7.2.6 float 和 char(NDRange-private)

本节在 kernel 采用 NDRange 编程模式，rt 变量采用 private 存储模式的情况下，分析比较 float 数据类型与 char 类型对 FPGA 实现的影响。通过 7.2.3 节的对比分析，NDRange 编程模式下，rt 变量采用 private 和 local 存储模式的效果相同。因此，在这里只分析 rt 变量采用 private 存储模式的情况。

1. kernel 代码对比

图 7-22(a)为采用 float 数据类型实现的 kernel 程序,图 7-22(b)为采用 char 数据类型实现的 kernel 程序。float 数据类型位数为 32 位,char 数据类型位数为 8 位。

```
__kernel void mnist_simple (__global const float *restrict dev_x, //28*28
                            __global const float *restrict dev_w, //784*10
                            __global const float *restrict dev_b,//10
                            __global float *restrict dev_y) //10
{
    float rt[10]={0.0};
    unsigned thread_id = get_global_id(0);
    for (int i=0;i<784;i++)
      {
         rt[thread_id]=rt[thread_id]+dev_x[i]*dev_w[thread_id+i*10];
      }
     dev_y[thread_id]=rt[thread_id]+dev_b[thread_id];
}
```

(a) float

```
__kernel void mnist_simple (__global const char *restrict dev_x, //28*28
                            __global const char *restrict dev_w, //784*10
                            __global const char *restrict dev_b,//10
                            __global char *restrict dev_y) //10
{
    char rt[10]={0};
    unsigned thread_id = get_global_id(0);
    for (int i=0;i<784;i++)
      {
         rt[thread_id]=rt[thread_id]+dev_x[i]*dev_w[thread_id+i*10];
      }
     dev_y[thread_id]= rt[thread_id]+dev_b[thread_id];
}
```

(b) char

图 7-22　kernel 代码对比

2. FPGA 资源使用估计对比

在 mnist_simple.log 中查看 FPGA 资源使用估计报告,图 7-23(a)为 floats 数据类型的 FPGA 资源使用估计信息,图 7-23(b)为 char 数据类型的 FPGA 资源使用估计信息。结果表明,char 数据类型使用的 FPGA 资源更少。

3. kernel 运行时间对比

图 7-24(a)给出了 float 数据类型的 101 次 kernel 的运行时间,图 7-24(b)给出了 char 数据类型的 101 次 kernel 的运行时间。结果对比,该情况下,char 数据类型的 kernel 运行时间更少。

```
+----------------------------------------------------+
; Estimated Resource Usage Summary                   ;
+--------------------------------+-------------------+
; Resource                       + Usage             ;
+--------------------------------+-------------------+
; Logic utilization              ;   24%             ;
; ALUTs                          ;   16%             ;
; Dedicated logic registers      ;    9%             ;
; Memory blocks                  ;   14%             ;
; DSP blocks                     ;    1%             ;
+----------------------------------------------------+
```

(a) float

```
+----------------------------------------------------+
; Estimated Resource Usage Summary                   ;
+--------------------------------+-------------------+
; Resource                       + Usage             ;
+--------------------------------+-------------------+
; Logic utilization              ;   23%             ;
; ALUTs                          ;   15%             ;
; Dedicated logic registers      ;    9%             ;
; Memory blocks                  ;   14%             ;
; DSP blocks                     ;    1%             ;
+----------------------------------------------------+
```

(b) char

图 7-23　FPGA 使用资源估计对比

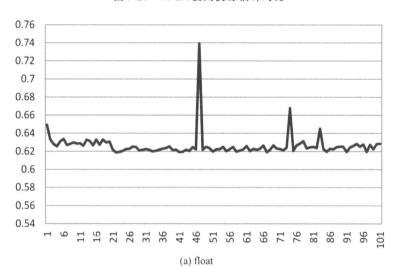

(a) float

(b) char

图 7-24　kernel 运行时间对比

4. 运行结果对比

图 7-25 为终端显示的 FPGA 运行结果对比,图 7-25(a)为 float 数据类型的运行结果,图 7-25(b)为 char 数据类型的运行结果。结果表明,char 数据类型略差于 float 数据类型。

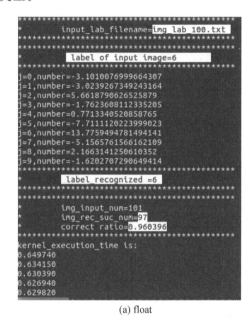

(a) float　　　　　　　　(b) char

图 7-25　FPGA 运行结果对比

小结 2:通过 7.2.5~7.2.6 节的对比分析可以发现,无论是 single work item 编程模式还是 NDRange 编程模式,char 数据类型在资源使用及 kernel 执行时间上都优于 float 数据类型,但是在识别成功率上略逊一等。

7.3　神经网络算法的 ARM 与 FPGA 实现方式对比

神经网络算法除了用 FPGA 硬件实现外,还可以用嵌入式处理器以软件代码的形式实现,DE10_nano 开发板的 FPGA 内嵌双核 ARM 处理器,也可以实现神经网络算法。下面比较分析这两种实现方式的特点。

7.3.1　ARM 和 FPGA(float 数据类型)

FPGA 实现选取 single work item-local-float 组合,ARM 实现时数据类型取 float。

1. kernel 代码对比

采用 OpenCL 实现方式,运行 host 程序时需要一个 FPGA 编程文件。ARM 实现神经网络算法时,神经网络算法的功能由 host 程序实现,kernel 程序的作用只是生成 aocx 文件,不实现任何功能,因此可编写一个空的 kernel 程序。图 7-26(a)是 ARM 实现方式时的 kernel 代码，是一个空函数。图 7-26(b) 是 FPGA 实现方式时的 kernel 代码，是 single work item-local-float 组合。

```
__kernel void mnist_simple( )
{

}
```

(a) ARM

```
__kernel void mnist_simple (__global const float *restrict dev_x, //28*28
                            __global const float *restrict dev_w, //784*10
                            __global const float *restrict dev_b,//10
                            __global float *restrict dev_y) //10
 {

  __local float rt[10];
  for (unsigned thread_id = 0;thread_id<10;thread_id++)
  {
   rt[thread_id]=0.0;
   for (int i=0;i<784;i++)
     {
       rt[thread_id]=rt[thread_id]+dev_x[i]*dev_w[thread_id+i*10];
     }
    dev_y[thread_id]=rt[thread_id]+dev_b[thread_id];
  }
}
```

(b) single work item-local-float

图 7-26 kernel 代码对比

2. FPGA 资源使用估计对比

在 mnist_simple.log 中查看 FPGA 资源使用估计报告，图 7-27(a) 为 ARM 实现方式的 FPGA 资源使用估计信息，图 7-27(b) 为 single work item-local-float 实现方式的 FPGA 资源使用估计信息。因为 ARM 实现方式的 FPGA 不实现任何逻辑功能，因此使用的 FPGA 资源更少。

```
+---------------------------------------------------------------------+
; Estimated Resource Usage Summary                                    ;
+----------------------------------------+----------------------------+
; Resource                               + Usage                      ;
+----------------------------------------+----------------------------+
; Logic utilization                      ;    5%                      ;
; ALUTs                                  ;    3%                      ;
; Dedicated logic registers              ;    2%                      ;
; Memory blocks                          ;    4%                      ;
; DSP blocks                             ;    0%                      ;
+----------------------------------------+----------------------------;
```

(a) ARM

图 7-27 FPGA 资源使用估计对比

```
+---------------------------------------------+
; Estimated Resource Usage Summary            ;
+----------------------------------+----------+
; Resource                         + Usage    ;
+----------------------------------+----------+
; Logic utilization                ;   25%    ;
; ALUTs                            ;   17%    ;
; Dedicated logic registers        ;   10%    ;
; Memory blocks                    ;   15%    ;
; DSP blocks                       ;    1%    ;
+---------------------------------------------+
```

(b) single work item-local-float

图 7-27 （续）

3. 神经网络算法运行时间对比

图 7-28(a)给出了 ARM 实现的 101 次神经网络算法的运行时间，图 7-28(b)给出了 single work item-local-float 实现的 101 次 kernel 的运行时间。对比结果可以发现，二者的算法运行时间基本相同。

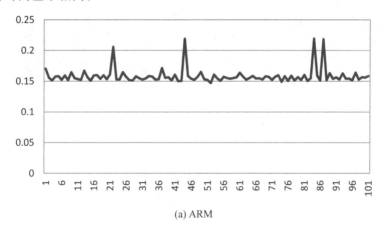

(a) ARM

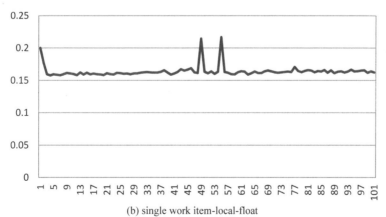

(b) single work item-local-float

图 7-28　神经网络算法运行时间对比

4. 运行结果对比

图 7-29 为终端显示的 FPGA 运行结果对比,图 7-29(a)为 ARM 实现的运行结果,图 7-29(b)为 single work item-local-float 实现的运行结果。结果表明,二者的分类效果相同。

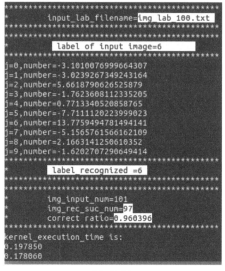
(a) ARM

(b) single work item-local-float

图 7-29　FPGA 运行结果对比

7.3.2　ARM 和 FPGA(char 数据类型)

FPGA 实现选取 single work item-local-char 组合,ARM 实现时的数据类型取 char。

1. kernel 代码对比

图 7-30(a)是 ARM 实现方式的 kernel 代码,是一个空函数。图 7-30(b)是 FPGA 实现方式的 kernel 代码,是 single work item-local-char 组合。

2. FPGA 资源使用估计对比

在 mnist_simple.log 中查看 FPGA 资源使用估计报告,图 7-31(a)为 ARM 实现方式的 FPGA 资源使用估计信息,图 7-31(b)为 single work item-local-char 实现方式的 FPGA 资源使用估计信息。因为 ARM 实现方式的 FPGA 不实现任何逻辑功能,因此使用的 FPGA 资源更少。

```
__kernel void mnist_simple(  )
{

}
```
(a) ARM

图 7-30　kernel 代码对比

```
__kernel void mnist_simple (__global const char *restrict dev_x, //28*28
                            __global const char *restrict dev_w, //784*10
                            __global const char *restrict dev_b,//10
                            __global char *restrict dev_y) //10
{
    __local char rt[10];
    for (unsigned thread_id=0;thread_id<10;thread_id++)
    {
        rt[thread_id]=0;
        for (int i=0;i<784;i++)
        {
            rt[thread_id]=rt[thread_id]+dev_x[i]*dev_w[thread_id+i*10];
        }
        dev_y[thread_id]= rt[thread_id]+dev_b[thread_id];
    }
}
```

(b) single work item-local-char

图 7-30 （续）

```
+-----------------------------------------------------------+
; Estimated Resource Usage Summary                          ;
+-----------------------------------+-----------------------+
; Resource                          ; + Usage               ;
+-----------------------------------+-----------------------+
; Logic utilization                 ;   5%                  ;
; ALUTs                             ;   3%                  ;
; Dedicated logic registers         ;   2%                  ;
; Memory blocks                     ;   4%                  ;
; DSP blocks                        ;   0%                  ;
+-----------------------------------+-----------------------+
```

(a) ARM

```
+-----------------------------------------------------------+
; Estimated Resource Usage Summary                          ;
+-----------------------------------+-----------------------+
; Resource                          ; + Usage               ;
+-----------------------------------+-----------------------+
; Logic utilization                 ;   22%                 ;
; ALUTs                             ;   15%                 ;
; Dedicated logic registers         ;   9%                  ;
; Memory blocks                     ;   15%                 ;
; DSP blocks                        ;   1%                  ;
+-----------------------------------+-----------------------+
```

(b) single work item-local-float

图 7-31 FPGA 资源使用估计对比

3. 神经网络算法运行时间对比

图 7-32(a)给出了 ARM 实现的 101 次神经网络算法的运行时间，图 7-32(b)给出了 single work item-local-char 实现的 101 次 kernel 的运行时间。结果表明，ARM 实现方式的算法运行时间更少。

4. 运行结果对比

图 7-33 为终端显示的 FPGA 运行结果对比，图 7-33(a)为 ARM 实现的运行结果，图 7-33(b)为 single work item-local-char 实现的运行结果。结果表明，二者的分类效果相同。

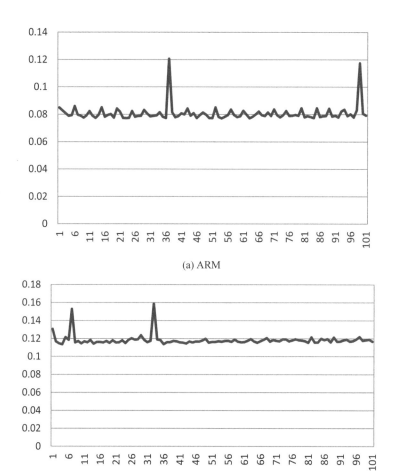

(a) ARM

(b) single work item-local-char

图 7-32 神经网络算法运行时间对比

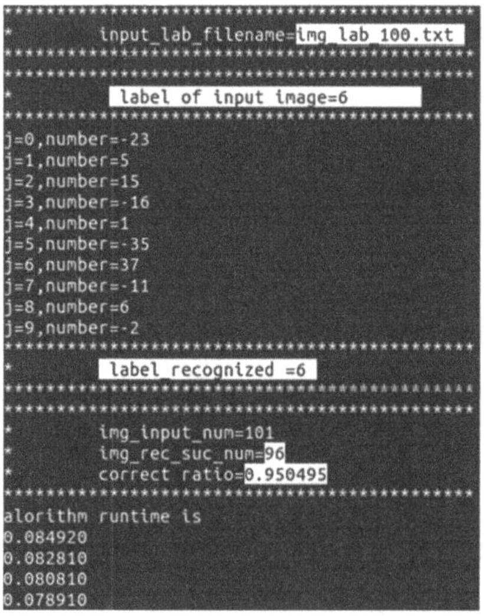

(a) ARM 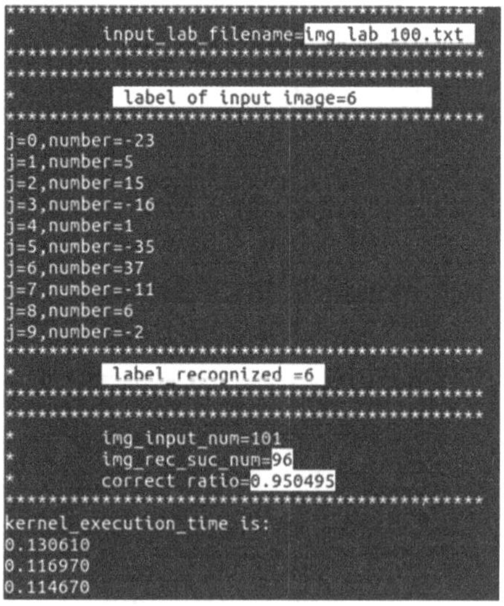 (b) single work item-local-char

图 7-33 FPGA 运行结果对比

7.4 host 代码与 kernel 的对应

1. single work item

```
status= clEnqueueNDRangeKernel(queue, kernel, 1, 0, GSize, WSize, 0, NULL,
&event_kernel);
```

clEnqueueNDRangeKernel 函数中的参数 GSize 和 WSize 设置如下：

```
static const size_t GSize[]={1};        //global size
static const size_t WSize[]={1};        //local size
```

2. NDRange

```
status= clEnqueueNDRangeKernel(queue, kernel, 1, 0, GSize, WSize, 0, NULL,
&event_kernel);
```

clEnqueueNDRangeKernel 函数中的参数 GSize 和 WSize 设置如下：

```
static const size_t GSize[]={10};   //global size
static const size_t WSize[]={1}; //local size
```

3. ARM 实现

ARM 实现时，kernel 代码为空函数，功能为产生 aocx 文件以实现 FPGA 编程，不实现任何功能，因此 host 代码中无须创建 kernel，无须为 kernel 端口分配内存，无须将 host 数据写入 kernel 内存，无须设置 kernel 参数，无须执行 kernel，无须将 kernel 内存中的数据读取到 host，但需要在 host 代码中实现神经网络算法。

习题 7

7.1 kernel 编程时可以尝试从哪些方面提高算法性能指标？
7.2 kernel 算法的性能指标主要包括哪些？
7.3 为什么基于 ARM 实现 kernel 算法时仍需要一个空的 kernel 函数？
7.4 kernel 程序中 float 类型数据的长度是多少？char 类型数据的长度是多少？
7.5 kernel 程序中定义的存储变量在存储类型缺省时代表什么类型？
7.6 kernel 程序中定义的 local 存储变量能否在定义时直接赋予初始值？
7.7 在 host 程序中如何设置才能满足 kernel 程序采用 single work item 或 NDRange 编程方式？
7.8 host(ARM)实现 kernel 程序功能时，host 代码有何特点？

第 8 章

具有一个隐形层的神经网络算法模型的 OpenCL 实现

本章介绍具有一个隐形层的神经网络模型的实现，在单层神经网络模型分析比较的基础上，进一步分析比较 kernel 函数个数对算法实现的影响，然后介绍 kernel 函数之间 pipe 和 channel 通信机制的使用方法。

8.1 一个隐形层的简易神经网络算法原理

具有一个隐形层的简易神经网络可以看作两个单层神经网络的级联。图 8-1 为含有一个隐形层的神经网络实现 MNIST 数据识别的示意图。本书选取隐形层的神经元个数为 100，即第 1 层神经网络的权值个数为 784×100，第 2 层神经元的个数为 10，神经网络的权值个数为 100×10。

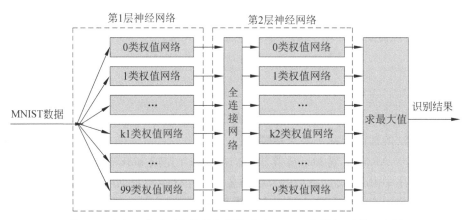

图 8-1 含有一个隐形层的神经网络实现 MNIST 数据识别的示意图

1. 第 1 层神经网络

图 8-2 为图 8-1 中第 1 层神经网络中的 k1（k1 取值为[0～99]）类权值网络的内部算

法实现原理。ReLU 为激活函数,其原理如图 8-3 所示。

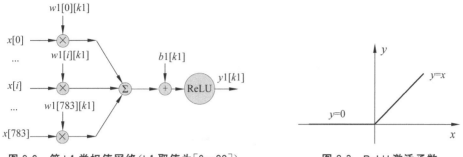

图 8-2　第 k1 类权值网络(k1 取值为[0~99])　　　图 8-3　ReLU 激活函数

第 1 层神经网络的权值个数为 784×100,在结构上为 784 行、100 列的数组。本书中,权值是训练好后存储在存储器中的。在读取数据时,为了方便算法实现,一般根据编号进行数据读取,权值数组的编号如图 8-4 所示。

w1[0]	w1[1]	w1[2]	…	w1[99]
w1[100]	w1[101]	w1[102]	…	w1[199]
…	…	…	…	…
w1[i*100]	w1[i*100+1]	w1[i*100+2]	…	w1[i*100+99]
…	…	…	…	…
w1[78300]	w1[78301]	w1[78302]	…	w1[78399]

图 8-4　权值 w1 的存储结构

根据第 1 层神经网络结构及图 8-4 所示的权值存储结构,第 1 层神经网络的算法描述如下。

```
for(int k1=0;k1<100;k1++)                //k1 对应图 8-4 中的列数
  {
    float y1[k1]=0.0;
    for (int i=0;i<784;i++)              //i 对应图 8-4 中的行数
      {
        y1[k1]=y1[k1]+x[i] * w1[k1+i * 100];
      }
    y1[k1]=y1[k1]+b1[k1];
    y1[k1]=(y1[k1]>0)? y1[k1]:0;         //ReLU
  }
```

2. 第 2 层神经网络

图 8-1 中的全连接网络代表第 1 层神经网络的 100 个输出数据全部连接到了第 2 层的 10 个神经元。图 8-5 为图 8-1 中第 2 层神经网络中的 k2(k2 取值为[0~9])类权值网络的内部算法实现原理图。

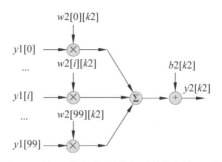

图 8-5 第 k2 类权值网络(k2 取值为[0~9])

第 2 层神经网络的权值个数为 100×10,在结构上为 100 行、10 列的数组。本书中,权值是训练好后存储在存储器中的。在数据读取时,为了方便算法实现,一般根据编号进行数据读取,权值数组的编号如图 8-6 所示。

w2[0]	w2[1]	w2[2]	...	w2[9]
w2[10]	w2[11]	w2[12]	...	w2[19]
...	...	...		...
w2[i*10]	w2[i*10+1]	w2[i*10+2]	...	w2[i*10+9]
...	...	...		...
w2[990]	w2[991]	w2[992]	...	w2[999]

图 8-6 权值 w2 的存储结构

根据第 2 层神经网络结构及图 8-6 所示的权值存储结构,第 2 层神经网络的算法描述如下。

```
for (int k2=0;k2<10;k2++)              //k2 对应图 8-6 中的列数
  {
    float y2[k2]=0.0;
    for (int k1=0;k1<100;k1++)         //k1 对应图 8-6 中的行数
      {
        y2[k2]=y2[k2]+y1[k1] * w2[k2+k1 * 10];
      }
    y2[k2]=y2[k2]+b2[k2];
}
```

8.2 具有一个隐形层的神经网络的 TensorFlow 实现及训练

根据 8.1 节的算法分析,建立算法的 TensorFlow 模型,代码如下。

```
#coding: utf-8
import tensorflow as tf
import numpy as np
```

```python
import sys
sys.path.append("/home/tensorflow/design/my_mnist/")

import input_data
#载入数据集
Mnist=input_data.read_data_sets("/home/tensorflow/design/my_mnist/MNIST_data",one_hot=True)

#每个批次的大小
batch_size=100
#计算一共有多少个批次
n_batch=mnist.train.num_examples //batch_size

in_units=784
h1_units=100
#定义两个placeholder
x=tf.placeholder(tf.float32,[None,in_units])
y=tf.placeholder(tf.float32,[None,10])

#创建一个简单的神经网络

W1=tf.Variable(tf.truncated_normal([in_units,h1_units],stddev=0.1))
b1=tf.Variable(tf.zeros([h1_units]))
W2=tf.Variable(tf.zeros([h1_units,10]))
b2=tf.Variable(tf.zeros([10]))
keep_prob=tf.placeholder(tf.float32)
hidden1=tf.nn.relu(tf.matmul(x,W1)+b1)
hidden1_drop=tf.nn.dropout(hidden1,keep_prob)
prediction=tf.nn.softmax(tf.matmul(hidden1_drop,W2)+b2)
#使用交叉熵
loss=
tf.reduce_mean(tf.nn.softmax_cross_entropy_with_logits(labels=y,logits=prediction))
#使用梯度下降法
train_step=tf.train.GradientDescentOptimizer(0.2).minimize(loss)
#初始化变量
init=tf.global_variables_initializer()

#结果存放在一个布尔型列表中,#argmax返回一维张量中最大值所在的位置
correct_prediction=tf.equal(tf.argmax(y,1),tf.argmax(prediction,1))
#求准确率
accuracy=tf.reduce_mean(tf.cast(correct_prediction,tf.float32))

with tf.Session() as sess:
```

```python
    sess.run(init)
    for epoch in range(101):
        for batch in range(n_batch):
            batch_xs,batch_ys=mnist.train.next_batch(batch_size)
            sess.run(train_step,feed_dict={x:batch_xs,y:batch_
            ys,keep_prob:0.5})
        acc_train=sess.run(accuracy,feed_dict={x:mnist.train.images,y:
mnist.train.labels,keep_prob:1.0})
        print("Iter "+str(epoch)+",Training Accuracy "+str(acc_train))
    acc=sess.run(accuracy,feed_dict={x:mnist.test.images,y:mnist.test.
labels,keep_prob:1.0})
    print("Iter "+str(epoch)+",Testing Accuracy "+str(acc))

    w1_sim_val, b1_sim_val,w2_sim_val, b2_sim_val=sess.run([W1,b1,W2,b2])
        print("text write sucessful")

    np.savetxt("w1_sim.txt", w1_sim_val.reshape(-1), fmt="%.31f",
delimiter=",")
    np.savetxt("b1_sim.txt", b1_sim_val.reshape(-1), fmt="%.31f",
delimiter=",")
    np.savetxt("w2_sim.txt", w2_sim_val.reshape(-1), fmt="%.31f",
delimiter=",")
    np.savetxt("b2_sim.txt", b2_sim_val.reshape(-1), fmt="%.31f",
delimiter=",")
```

训练结果如图 8-7 所示。

```
Iter 81,Training Accuracy 0.97505456
Iter 82,Training Accuracy 0.97467273
Iter 83,Training Accuracy 0.9751273
Iter 84,Training Accuracy 0.97574544
Iter 85,Training Accuracy 0.9752
Iter 86,Training Accuracy 0.9757636
Iter 87,Training Accuracy 0.9759091
Iter 88,Training Accuracy 0.9765818
Iter 89,Training Accuracy 0.9762909
Iter 90,Training Accuracy 0.9764182
Iter 91,Training Accuracy 0.9767454
Iter 92,Training Accuracy 0.9768
Iter 93,Training Accuracy 0.9768909
Iter 94,Training Accuracy 0.97678185
Iter 95,Training Accuracy 0.9771818
Iter 96,Training Accuracy 0.97743636
Iter 97,Training Accuracy 0.9775091
Iter 98,Training Accuracy 0.9773273
Iter 99,Training Accuracy 0.9771818
Iter 100,Training Accuracy 0.97745454
Iter 100,Testing Accuracy 0.9697
text write sucessful
```

图 8-7　TensorFlow 模型运行结果

8.3 具有一个隐形层的神经网络算法的 OpenCL 实现

8.3.1 ARM 实现

与 7.3 节相同,具有一个隐形层的神经网络算法也可以通过 ARM 软件实现。

1. kernel 代码

图 8-8 为 ARM 实现方式的 kernel 代码,kernel 程序的作用只是生成 aocx 文件,不实现任何功能。

```
__kernel void mnist_hidden (  )
 {

 }
```

图 8-8 ARM 实现的 kernel 代码

2. FPGA 资源使用估计

在 mnist_hidden.log 中查看 FPGA 资源使用估计报告,图 8-9 为 ARM 实现方式的 FPGA 资源使用估计信息。

```
+----------------------------------------------------+
; Estimated Resource Usage Summary                   ;
+----------------------------+-----------------------+
; Resource                   + Usage                 ;
+----------------------------+-----------------------+
; Logic utilization          ;   5%                  ;
; ALUTs                      ;   3%                  ;
; Dedicated logic registers  ;   2%                  ;
; Memory blocks              ;   4%                  ;
; DSP blocks                 ;   0%                  ;
+----------------------------+-----------------------+
```

图 8-9 FPGA 资源使用估计

3. 神经网络算法运行时间

图 8-10 给出了 ARM 实现的 101 次神经网络算法的运行时间(ms)。

4. 运行结果

图 8-11 为终端显示的 FPGA 运行结果,与第 7 章的单层神经网络相比,其成功识别的图片数量从 97 提高到 100。

8.3.2 single work item 格式,一个 kernel

本节的 kernel 代码采用 single work item-local-float 组合实现,并将两层神经网络放到同一个 kernel 代码中。

1. kernel 代码

图 8-12 为 single work item-local-float 组合实现具有一个隐形层神经网络算法的

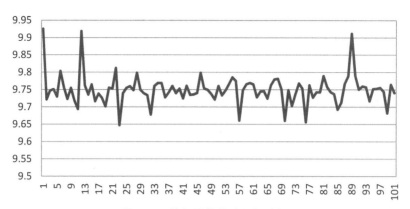

图 8-10 神经网络算法运行时间

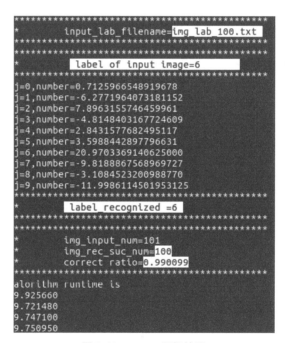

图 8-11 FPGA 运行结果

kernel 代码。

2. FPGA 资源使用估计

在 mnist_hidden.log 中查看 FPGA 资源使用估计报告，图 8-13 为当前实现方式的 FPGA 资源使用估计信息。

3. kernel 运行时间

图 8-14 给出了当前实现方式的 101 次神经网络算法的运行时间（ms）。从图 8-14 中可以发现，kernel 运行时间的差异较大，大概在 20～200ms 之间波动。

第8章 具有一个隐形层的神经网络算法模型的OpenCL实现

```
__kernel void mnist_hidden_one_kernel (__global const float *restrict dev_x, //28*28
                          __global const float *restrict dev_w1, //784*100
                          __global const float *restrict dev_b1,//100
                          __global const float *restrict dev_w2, //100*10
                          __global const float *restrict dev_b2,//10
                          __global float *restrict dev_y2) //10

{
    __local float rt1[100];
    __local float rt2[10];
    __local float y1[100];

    for (int k=0;k<10;k++)
      {
        rt2[k]=0.0;
        #pragma unrool 1
        for (int thread_id=0;thread_id<100;thread_id++)
          {
            rt1[thread_id]=0.0;
            #pragma unrool 1
            for (int i=0;i<784;i++)
              {
                rt1[thread_id]=rt1[thread_id]+dev_x[i]*dev_w1[thread_id+i*100];
              }

            y1[thread_id]=rt1[thread_id]+dev_b1[thread_id];
            y1[thread_id]=(y1[thread_id]>0)?y1[thread_id]:0;
            rt2[k]=rt2[k]+y1[thread_id]*dev_w2[k+thread_id*10];
            dev_y2[k]=rt2[k]+dev_b2[k];
          }
      }
}
```

图 8-12　kernel 代码

```
+----------------------------------------------------+
; Estimated Resource Usage Summary                   ;
+-------------------------------+--------------------+
; Resource                      + Usage              ;
+-------------------------------+--------------------+
; Logic utilization             ;    21%             ;
; ALUTs                         ;    13%             ;
; Dedicated logic registers     ;     9%             ;
; Memory blocks                 ;    22%             ;
; DSP blocks                    ;     2%             ;
+-------------------------------+--------------------+
```

图 8-13　FPGA 资源使用估计

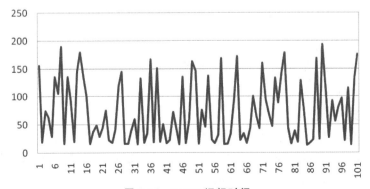

图 8-14　kernel 运行时间

4. 运行结果

图 8-15 为终端显示的 FPGA 运行结果，与第 7 章的单层神经网络相比，其成功识别的图片数量从 97 提高到 101。

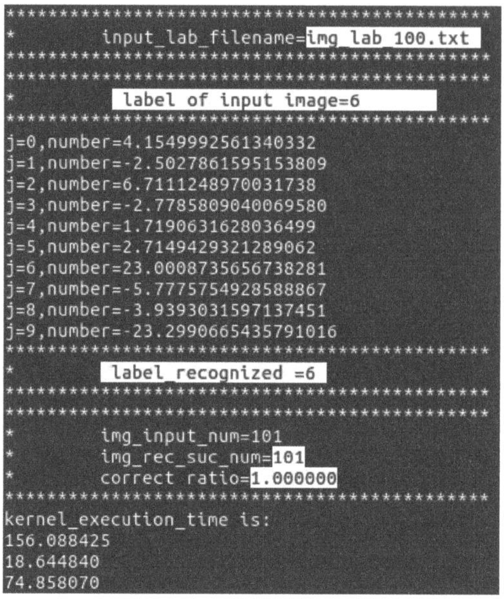

图 8-15　FPGA 运行结果

8.3.3　NDRange 格式，一个 kernel

本节的 kernel 代码采用 NDRange-local-float 组合实现，并将两层神经网络放到同一个 kernel 代码中。

1. kernel 代码

图 8-16 为 NDRange-local-float 组合实现具有一个隐形层神经网络算法的 kernel 代码。

注意：该 kernel 代码使用了 barrier 函数，因为在 NDRange 模式下，不同维度的 work item 的执行顺序无法保障，因此使用 barrier 函数进行 work item 同步。

2. FPGA 资源使用估计

在 mnist_hidden.log 中查看 FPGA 资源使用估计报告，图 8-17 为当前实现方式的 FPGA 资源使用估计信息。

3. kernel 运行时间

图 8-18 给出了当前实现方式的 101 次神经网络算法的运行时间。从图 8-18 中可以发现，每次 kernel 的运行时间耗时较长，大概为 650.78ms。耗时较长的原因是因为使用 barrier 函数会限制 work item 的并行执行。

```
__kernel void mnist_hidden_one_kernel (__global const float *restrict dev_x, //28*28
                        __global const float *restrict dev_w1, //784*100
                        __global const float *restrict dev_b1,//100
                        __global const float *restrict dev_w2, //100*10
                        __global const float *restrict dev_b2,//10
                        __global float *restrict dev_y2) //10

{
    __local float rt1[100];
    __local float rt2[10];
    __local float dev_y1[100];
    unsigned thread_id1 = get_global_id(0);
    unsigned thread_id2 = get_global_id(1);
    {
    rt1[thread_id1]=0.0;
    for (int i=0;i<784;i++)
        {
            rt1[thread_id1]=rt1[thread_id1]+dev_x[i]*dev_w1[thread_id1+i*100];
        }
    dev_y1[thread_id1]=rt1[thread_id1]+dev_b1[thread_id1];
    dev_y1[thread_id1]=(dev_y1[thread_id1]>0)?dev_y1[thread_id1]:0;
    }
    barrier(CLK_LOCAL_MEM_FENCE);
{
    rt2[thread_id2]=0.0;
    {
        for (int i=0;i<100;i++)
        {
            rt2[thread_id2]=rt2[thread_id2]+dev_y1[i]*dev_w2[thread_id2+i*10];
        }
        dev_y2[thread_id2]=rt2[thread_id2]+dev_b2[thread_id2];
    }
}
    barrier(CLK_LOCAL_MEM_FENCE);
}
```

图 8-16 kernel 代码

```
+-----------------------------------------------+
; Estimated Resource Usage Summary              ;
+-----------------------------------+-----------+
; Resource                          + Usage     ;
+-----------------------------------+-----------+
; Logic utilization                 ;   29%     ;
; ALUTs                             ;   18%     ;
; Dedicated logic registers         ;   12%     ;
; Memory blocks                     ;   39%     ;
; DSP blocks                        ;    2%     ;
+-----------------------------------+-----------+
```

图 8-17 FPGA 资源使用估计

4. 运行结果

图 8-19 为终端显示的 FPGA 运行结果，与第 7 章的单层神经网络相比，其成功识别的图片数量从 97 提高到 101。

图 8-18 kernel 运行时间

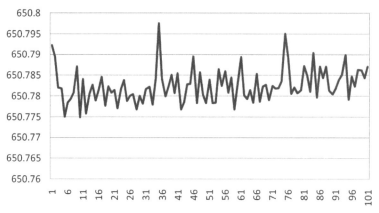

图 8-19 FPGA 运行结果

8.3.4 single work item 格式，两个 kernel

本节的 kernel 代码采用 single work item-local-float 组合实现，并用两个 kernel 函数分别实现神经网络算法，每层神经网络对应一个 kernel 函数。

1. kernel 代码

图 8-20 为 single work item-local-float 组合实现具有一个隐形层神经网络算法的 kernel 代码。该 kernel 代码中含有两个 kernel 函数，即 mnist_hidden_l1() 和 mnist_hidden_l2()，分别对应第 1 层神经网络和第 2 层神经网络。

```
// single work item kernel version
__kernel void mnist_hidden_l1 (__global const float *restrict dev_x, //28*28
                               __global const float *restrict dev_w1, //784*100
                               __global const float *restrict dev_b1,//100
                               __global float *restrict dev_y1 )//100
 {

   __local float rt[100];
  for (int thread_id=0;thread_id<100;thread_id++)
    {
       rt[thread_id]=0.0;
    for (int i=0;i<784;i++)
      {
         rt[thread_id]=rt[thread_id]+dev_x[i]*dev_w1[thread_id+i*100];
      }

    dev_y1[thread_id]=rt[thread_id]+dev_b1[thread_id];
    dev_y1[thread_id]=(dev_y1[thread_id]>0)?dev_y1[thread_id]:0;
   }

 }

__kernel void mnist_hidden_l2 ( __global float *restrict dev_y1,//100
                                __global const float *restrict dev_w2, //100*10
                                __global const float *restrict dev_b2,//10
                                __global float *restrict dev_y2) //10
    {

   __local float rt[10];
  for (int thread_id=0;thread_id<10;thread_id++)
    {
   rt[thread_id]=0.0;

    for (int i=0;i<100;i++)
      {
         rt[thread_id]=rt[thread_id]+dev_y1[i]*dev_w2[thread_id+i*10];
      }
     dev_y2[thread_id]=rt[thread_id]+dev_b2[thread_id];
    }
 }
```

图 8-20　kernel 代码

2. FPGA 资源使用估计

在 mnist_hidden.log 中查看 FPGA 资源使用估计报告,图 8-21 为当前实现方式的 FPGA 资源使用估计信息。与图 8-13 相比,两个 kernel 函数的形式比一个 kernel 函数的形式使用的 FPGA 资源更多。

```
+----------------------------------------------------------+
; Estimated Resource Usage Summary                         ;
+------------------------------------+---------------------+
; Resource                           + Usage               ;
+------------------------------------+---------------------+
; Logic utilization                  ;    25%              ;
; ALUTs                              ;    15%              ;
; Dedicated logic registers          ;    11%              ;
; Memory blocks                      ;    27%              ;
; DSP blocks                         ;     2%              ;
+------------------------------------+---------------------;
```

图 8-21　FPGA 资源使用估计

3. kernel 运行时间

图 8-22 给出了当前实现方式的 101 次神经网络算法的运行时间。与图 8-14 相比，其运行时间大幅减少。

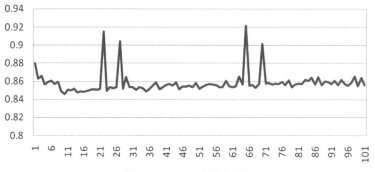

图 8-22　kernel 运行时间

4. 运行结果

图 8-23 为终端显示的 FPGA 运行结果，与图 8-15 对比，其识别成功图片的数目减少了一张。

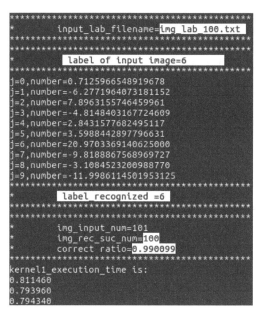

图 8-23　FPGA 运行结果

8.3.5　NDRange 格式，两个 kernel

本节的 kernel 代码采用 NDRange-local-float 组合实现，并用两个 kernel 函数分别实现神经网络算法，每层神经网络对应一个 kernel 函数。

1. kernel 代码

图 8-24 为 NDRange-local-float 组合实现具有一个隐形层神经网络算法的 kernel 代码。该 kernel 代码中含有两个 kernel 函数，即 mnist_hidden_l1() 和 mnist_hidden_l2()，分别对应第 1 层神经网络和第 2 层神经网络。

```
__kernel void mnist_hidden_l1 ( __global const float *restrict dev_x, //28*28
                                __global const float *restrict dev_w1, //784*100
                                __global const float *restrict dev_b1,//100
                                __global float *restrict dev_y1))//100
{

    __local float rt[100];

    unsigned thread_id = get_global_id(0);
    {
    rt[thread_id]=0.0;
    for (int i=0;i<784;i++)
      {
        rt[thread_id]=rt[thread_id]+dev_x[i]*dev_w1[thread_id+i*100];
      }
      dev_y1[thread_id]=rt[thread_id]+dev_b1[thread_id];
      dev_y1[thread_id]=(dev_y1[thread_id]>0)?dev_y1[thread_id]:0;
    }
}

__kernel void mnist_hidden_l2 ( __global float *restrict dev_y1,//100
                                __global const float *restrict dev_w2, //100*10
                                __global const float *restrict dev_b2,//10
                                __global float *restrict dev_y2) //10
{

    __local float rt[10];

    unsigned thread_id = get_global_id(0);
    rt[thread_id]=0.0;
    {
    for (int i=0;i<100;i++)
      {
        rt[thread_id]=rt[thread_id]+dev_y1[i]*dev_w2[thread_id+i*10];
      }
      dev_y2[thread_id]=rt[thread_id]+dev_b2[thread_id];
    }
}
```

图 8-24　kernel 代码

2. FPGA 资源使用估计

在 mnist_hidden.log 中查看 FPGA 资源使用估计报告，图 8-25 为当前实现方式的 FPGA 资源使用估计信息。与图 8-17 相比，两个 kernel 函数的形式比一个 kernel 函数的形式使用的 FPGA 资源更少。这个结论与 single work item-local-float 实现方式恰好相反。

3. kernel 运行时间

图 8-26 给出了当前实现方式的 101 次神经网络算法的运行时间。与图 8-18 相比，其

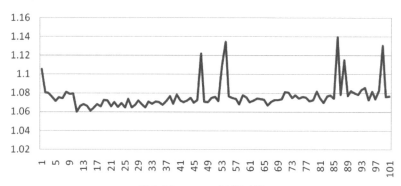

图 8-25 FPGA 资源使用估计

运行时间大幅减少。

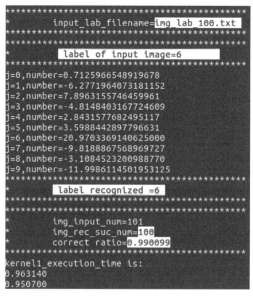

图 8-26 kernel 运行时间

4. 运行结果

图 8-27 为终端显示的 FPGA 运行结果,与图 8-19 对比,其识别成功的图片数目减少了一张。

图 8-27 FPGA 运行结果

8.3.6 single work item 格式，两个 kernel，channel

channel 功能可以让 kernel 函数直接通过 FIFO 进行数据交换，而不需要经过 host 处理器。本节的 kernel 代码采用 single work item-local-float 组合实现，并用两个 kernel 函数分别实现神经网络算法，每层神经网络对应一个 kernel 函数。与 8.3.4 节的不同之处在于，本节中的 kernel 函数之间采用 channel 通信机制。

1. kernel 代码

图 8-28 为 single work item-local-float 组合实现具有一个隐形层神经网络算法的 kernel 代码。该 kernel 代码中含有两个 kernel 函数，即 mnist_hidden_l1() 和 mnist_hidden_l2()，分别对应第 1 层神经网络和第 2 层神经网络。

```
#pragma OPENCL EXTENSION cl_intel_channels : enable
channel float dev_y1 __attribute__((depth(100))) ;

__kernel void mnist_hidden_l1 (__global const float *restrict dev_x, //28*28
                               __global const float *restrict dev_w1, //784*100
                               __global const float *restrict dev_b1)//100

  {
    __local float rt[100];
    //thread_id = get_global_id(0);
   for (int thread_id=0;thread_id<100;thread_id++)
     {
      rt[thread_id]=0.0;
      for (int i=0;i<784;i++)
        {
           rt[thread_id]=rt[thread_id]+dev_x[i]*dev_w1[thread_id+i*100];
        }

        rt[thread_id]=rt[thread_id]+dev_b1[thread_id];
        rt[thread_id]=(rt[thread_id]>0)?rt[thread_id]:0;
        write_channel_intel(dev_y1, rt[thread_id]);
     }
  }


__kernel void mnist_hidden_l2 (
                               __global const float *restrict dev_w2, //100*10
                               __global const float *restrict dev_b2,//10
                               __global float *restrict dev_y2) //10
   {
    __local float rt[10];
    __local float hidden_buf[100];

     for (int i=0;i<100;i++)
       {
         hidden_buf[i]=read_channel_intel(dev_y1);
       }
    for (int thread_id=0;thread_id<10;thread_id++)
      {
        rt[thread_id]=0.0;

        for (int i=0;i<100;i++)
          {
            rt[thread_id]=rt[thread_id]+hidden_buf[i]*dev_w2[thread_id+i*10];
          }
         dev_y2[thread_id]=rt[thread_id]+dev_b2[thread_id];
      }
   }
```

图 8-28　kernel 代码

代码♯pragma OPENCL EXTENSION cl_intel_channels : enable 的功能是使 channel 具有扩展功能。代码 channel float dev_y1_ _attribute_ _((depth(100)))的功能是定义一个深度为 100、位宽为 32 位浮点数、名称为 dev_y1 的 channel。代码 write_channel_intel(dev_y1,rt[thread_id])和 hidden_buf[i]=read_channel_intel(dev_y1)分别实现写 channel 和读 channel 的功能。

2. FPGA 资源使用估计

在 mnist_hidden.log 中查看 FPGA 资源使用估计报告,图 8-29 为当前实现方式的 FPGA 资源使用估计信息。与图 8-21 相比,使用 channel 实现方式使用的 FPGA 资源更少。

```
+---------------------------------------------------+
; Estimated Resource Usage Summary                  ;
;--------------------------------+------------------;
; Resource                       + Usage            ;
;--------------------------------+------------------;
; Logic utilization              ;   23%            ;
; ALUTs                          ;   15%            ;
; Dedicated logic registers      ;   10%            ;
; Memory blocks                  ;   22%            ;
; DSP blocks                     ;    2%            ;
+---------------------------------------------------+
```

图 8-29　FPGA 资源使用估计

3. kernel 运行时间

图 8-30 给出了当前实现方式的 101 次神经网络算法的运行时间。与图 8-22 相比,kernel 运行时间略有减少。

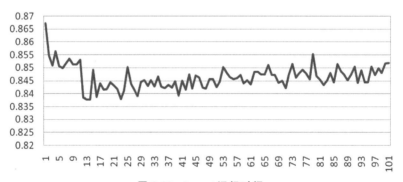

图 8-30　kernel 运行时间

4. 运行结果

图 8-31 为终端显示的 FPGA 运行结果,与图 8-23 对比,其识别成功的图片数目相同。

8.3.7　single work item 格式,两个 kernel,pipe

与 channel 功能类似,pipe 功能也可以让 kernel 函数直接通过 FIFO 进行数据交换,而不需要经过 host 处理器。本节的 kernel 代码采用 single work item-local-float 组合实

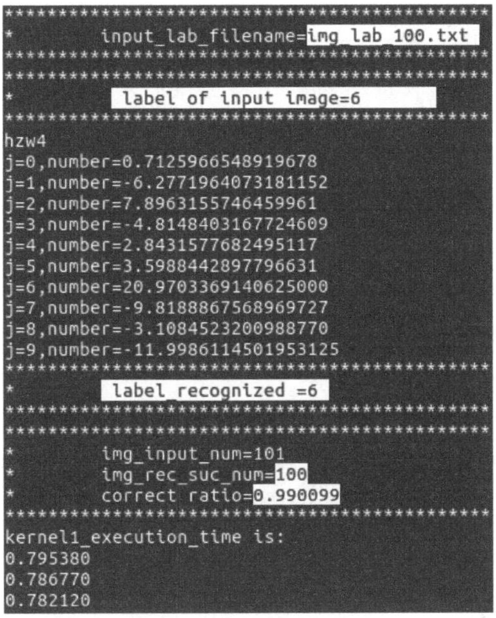

图 8-31 FPGA 运行结果

现,并用两个 kernel 函数分别实现神经网络算法,每层神经网络对应一个 kernel 函数。与 8.3.5 节的不同之处在于,本节中的 kernel 函数之间采用 pipe 通信机制。

1. kernel 代码

图 8-32 为 single work item-local-float 组合实现具有一个隐形层神经网络算法的 kernel 代码。该 kernel 代码中含两个 kernel 函数,即 mnist_hidden_l1() 和 mnist_hidden_l2(),分别对应第 1 层神经网络和第 2 层神经网络。

在此通过图 8-32 中的 kernel 代码引入 pipe 功能的使用方法。

(1) write_pipe 的定义。

例如:kernel 函数 mnist_hidden_l1() 的最后一个参数的定义为

```
write_only pipe float __attribute__ ((depth(100))) dev_y1
```

(2) read_pipe 的定义。

例如:kernel 函数 mnist_hidden_l2() 的第一个参数的定义为

```
read_only pipe float __attribute__ ((depth(100))) dev_y1
```

连接在一起的 pipe 要求名称一致。

(3) 写 pipe。

例如代码:

```
write_pipe(dev_y1,&rt[thread_id])
```

```
__kernel void mnist_hidden_l1 (__global const float *restrict dev_x, //28*28
                               __global const float *restrict dev_w1, //784*100
                               __global const float *restrict dev_b1,//100
                               write_only pipe float __attribute__((depth(100))) dev_y1)//100
{
    __local  float rt[100];
  for (int thread_id=0;thread_id<100;thread_id++)
    {
     rt[thread_id]=0.0;

      for (int i=0;i<784;i++)
       {
          rt[thread_id]=rt[thread_id]+dev_x[i]*dev_w1[thread_id+i*100];
       }
      rt[thread_id]=rt[thread_id]+dev_b1[thread_id];
      rt[thread_id]=(rt[thread_id]>0)?rt[thread_id]:0;
      write_pipe(dev_y1, &rt[thread_id]);
    }
}
__kernel void mnist_hidden_l2 ( read_only pipe float __attribute__((depth(100))) dev_y1,//100
                                __global const float *restrict dev_w2, //100*10
                                __global const float *restrict dev_b2,//10
                                __global float *restrict dev_y2) //10
    {
    __local float rt[10] ;
    __local float hidden_buf[100];
   for (int i=0;i<100;i++)
      {
         hidden_buf[i]=0.0;
         read_pipe(dev_y1, &hidden_buf[i]);
      }

   for (int thread_id=0;thread_id<10;thread_id++)
     {
      rt[thread_id]=0.0;
        for (int i=0;i<100;i++)
         {
           rt[thread_id]=rt[thread_id]+hidden_buf[i]*dev_w2[thread_id+i*10];
         }
        dev_y2[thread_id]=rt[thread_id]+dev_b2[thread_id];
     }
  }
```

图 8-32　kernel 代码

(4) 读 pipe。

例如代码：

```
read_pipe(dev_y1,&hidden_buffer[i])
```

2. FPGA 资源使用估计

在 mnist_hidden.log 中查看 FPGA 资源使用估计报告，图 8-33 为当前实现方式的 FPGA 资源使用估计信息。与图 8-29 相比，channel 实现方式与 pipe 实现方式使用的 FPGA 资源相同。

3. kernel 运行时间

图 8-34 给出了当前实现方式的 101 次神经网络算法的运行时间。与图 8-30 相比，

```
+-------------------------------------------+--------+
; Estimated Resource Usage Summary                   ;
+-------------------------------------------+--------+
; Resource                                  + Usage  ;
+-------------------------------------------+--------+
; Logic utilization                         ;  23%   ;
; ALUTs                                     ;  15%   ;
; Dedicated logic registers                 ;  10%   ;
; Memory blocks                             ;  22%   ;
; DSP blocks                                ;  2%    ;
+-------------------------------------------+--------;
```

图 8-33　FPGA 资源使用估计

kernel 的运行时间基本相同。

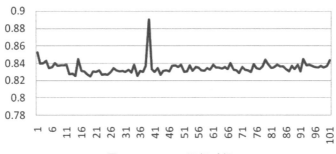

图 8-34　kernel 运行时间

4. 运行结果

图 8-35 为终端显示的 FPGA 运行结果，与图 8-31 对比，其识别成功的图片数目相同。

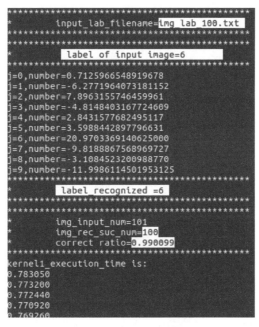

图 8-35　FPGA 运行结果

习题 8

8.1 一个 kernel 实现方式和多个 kernel 实现方式在代码形式上有何区别?

8.2 在 NDRange 编程模式下,不同维度的 work item 的执行顺序如何实现同步?

8.3 简述 channel 通信机制的原理。

8.4 如何实现 channel 功能? 如何实现 channel 的读写功能?

8.5 简述 pipe 通信机制的原理。

8.6 简述 write_pipe 和 read_pipe 的定义及读写功能的实现。

第 9 章

简易卷积神经网络的 OpenCL 实现

本章介绍基于 DE10_nano 开发板的简易卷积神经网络的 OpenCL 实现,为实现复杂的卷积神经网络奠定基础。

9.1 简易卷积神经网络算法结构与原理

由于 DE10_nano 开发板的片上资源有限,在实现复杂的神经网络算法时可能存在资源不足的问题,因此本书选择简易卷积神经网络举例说明。

如图 9-1 所示,本书采用的简易卷积神经网络包含一个卷积层、一个池化层和两个全连接层。

图 9-1 简易卷积神经网络结构

1. 卷积层

卷积层使用 4 个 3×3×1 的卷积核,卷积步长为 1,padding(边界处理)方式为通过补零使卷积的输入和输出保持相同尺寸,激活函数为 ReLU。MNIST 数据格式为 28×28,为了保持卷积运算前后的图像尺寸一致,在进行卷积前需要对边界进行处理,增加两行和两列,使数据格式变为 30×30,如图 9-2 所示。图 9-2 中的数字代表像素点编号,也可以看作是像素点的存储地址编号,若用数组表示整个像素点组合,则可以用数字作为数组脚标对像素点进行访问。

采用 4 个 3×3×1 卷积核可以理解为 4 个通道独立进行卷积运算。在每个通道内,卷积核的尺寸为 3×3×1。

例如在第 1 个通道内,当卷积步长为 1 时,与卷积核进行卷积运算的输入数据的顺序(从左到右,从上到下)及格式如图 9-3 所示。每行和每列都有 28 组 3×3×1 的输入数据,与卷积核进行卷积运算后,得到的输出数据格式为 28×28。因此,4 个通道的输出数

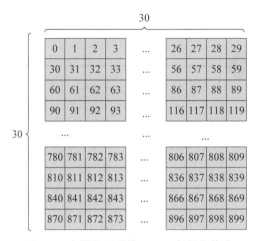

图 9-2 边界处理后的 MNIST 数据集格式

据的格式为 $28\times28\times4$。

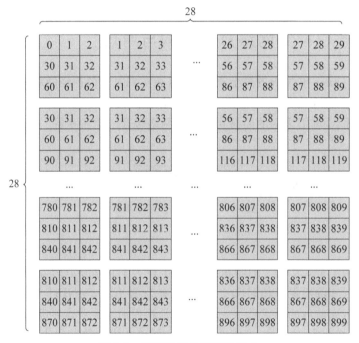

图 9-3 卷积运算的顺序及格式

2. 池化层

池化层又称降采样层,池化尺寸为 2×2,池化步长为 2,池化方式为最大池化,取池化窗口的最大值为池化输出。如图 9-4 所示,当只考虑一个通道时,卷积层的输出格式为 28×28,在相邻 4 组数据(虚线框内)进行最大池化操作,池化步长为 2,数据之间没有重叠,经过池化操作后,数据格式变为 14×14。因此,4 个通道的输出数据的格式为 $14\times14\times4$。

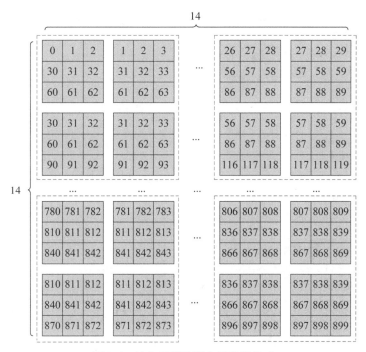

图 9-4 池化运算的顺序及输出格式

3. 全连接层 1

全连接层 1 使用了 50 个神经元，激活函数是 ReLU。全连接层 1 的实现原理与第 7 章和第 8 章的简易神经网络的实现原理相同，因此具体的实现原理不再详述。全连接层 1 的输入数据来自池化层，因此输入数据的格式为 $14 \times 14 \times 4$，因此全连接层 1 的连线数为 $14 \times 14 \times 4 \times 50$。

4. 全连接层 2

全连接层 2 使用了 10 个神经元，激活函数为 softmax，用于输出结果。全连接层 2 的输入数据来自全连接层 1，输入数据的格式为 1×50，因此全连接层 1 的连线数为 50×10。

9.2 简易卷积神经网络的 TensorFlow 实现及训练

1. 代码

实现代码如下。

```
#coding=utf-8
#---------------------
import input_data
import tensorflow as tf
import pylab
#读取数据
```

```python
mnist=input_data.read_data_sets('MNIST_data', one_hot=True)
#每个批次的大小
batch_size=100
#计算一共有多少个批次
n_batch=mnist.train.num_examples //batch_size
n_batch=10
#构建cnn网络结构
#自定义卷积函数
def conv2d(x,w):
    return tf.nn.conv2d(x,w,strides=[1,1,1,1],padding='SAME')
#自定义池化函数
def max_pool_2x2(x):
    return tf.nn.max_pool(x,ksize=[1,2,2,1],strides=[1,2,2,1],padding='SAME')
#设置占位符,尺寸为样本输入和输出的尺寸
x=tf.placeholder(tf.float32,[None,784])
y_=tf.placeholder(tf.float32,[None,10])
x_img=tf.reshape(x,[-1,28,28,1])

#设置第一个卷积层和池化层

w_conv1=tf.Variable(tf.truncated_normal([3,3,1,4],stddev=0.1))
b_conv1=tf.Variable(tf.constant(0.1,shape=[4]))
h_conv1=tf.nn.relu(conv2d(x_img,w_conv1)+b_conv1)
h_pool1=max_pool_2x2(h_conv1)

#设置第一个全连接层
w_fc1=tf.Variable(tf.truncated_normal([14*14*4,50],stddev=0.1))
b_fc1=tf.Variable(tf.constant(0.1,shape=[50]))
h_pool1_flat=tf.reshape(h_pool1,[-1,14*14*4])
h_fc1=tf.nn.relu(tf.matmul(h_pool1_flat,w_fc1)+b_fc1)
h_fc1_drop=tf.nn.dropout(h_fc1,0.5)

#设置第二个全连接层
w_fc2=tf.Variable(tf.truncated_normal([50,10],stddev=0.1))
b_fc2=tf.Variable(tf.constant(0.1,shape=[10]))
y_out=tf.nn.softmax(tf.matmul(h_fc1_drop,w_fc2)+b_fc2)

#使用交叉熵,梯度下降更有效
loss=tf.reduce_mean(tf.nn.softmax_cross_entropy_with_logits(labels=y_,
logits=y_out))
#使用梯度下降法
train_step=tf.train.GradientDescentOptimizer(0.2).minimize(loss)

#定义初始化操作
```

```python
init=tf.global_variables_initializer()
#建立正确率计算表达式
correct_prediction=tf.equal(tf.argmax(y_out,1),tf.argmax(y_,1))
accuracy=tf.reduce_mean(tf.cast(correct_prediction,tf.float32))
#开始输入数据,训练
with tf.Session() as sess:
    sess.run(init)
        for epoch in range(251):
            for batch in range(n_batch):
                batch_xs,batch_ys=mnist.train.next_batch(batch_size)
                sess.run(train_step,feed_dict={x:batch_xs,y_:batch_ys})
        train_acc=sess.run(accuracy,feed_dict={x:mnist.train.images,y_:mnist.train.labels})
        print("Iter "+str(epoch)+",Training Accuracy "+str(train_acc))
    test_acc=sess.run(accuracy,feed_dict={x:mnist.test.images,y_:mnist.test.labels})
    print("Testing Accuracy "+str(test_acc))

    w_conv1_val,b_conv1_val=sess.run([w_conv1,b_conv1])
    w_fc1_val,b_fc1_val=sess.run([w_fc1,b_fc1])
    w_fc2_val,b_fc2_val=sess.run([w_fc2,b_fc2])

    import numpy as np
    np.savetxt("./conv_txt/w_conv1.txt",w_conv1_val.reshape(-1),fmt="%f",delimiter=",")
    np.savetxt("./conv_txt/b_conv1.txt",b_conv1_val.reshape(-1),fmt="%f",delimiter=",")
np.savetxt("./conv_txt/w_fc1.txt",w_fc1_val.reshape(-1),fmt="%f",delimiter=",")
    np.savetxt("./conv_txt/b_fc1.txt",b_fc1_val.reshape(-1),fmt="%f",delimiter=",")
    np.savetxt("./conv_txt/w_fc2.txt",w_fc2_val.reshape(-1),fmt="%f",delimiter=",")
    np.savetxt("./conv_txt/b_fc2.txt",b_fc2_val.reshape(-1),fmt="%f",delimiter=",")
print("text write sucessful")
```

2. 运行结果

运行结果如图 9-5 所示。

3. 数据格式

为了在 FPGA 上运行神经网络,需要将 TensorFlow 训练后的卷积神经网络模型的权值和偏置保存为 txt 文件,因此需要明确数据的存储格式,根据数据的存储格式编写

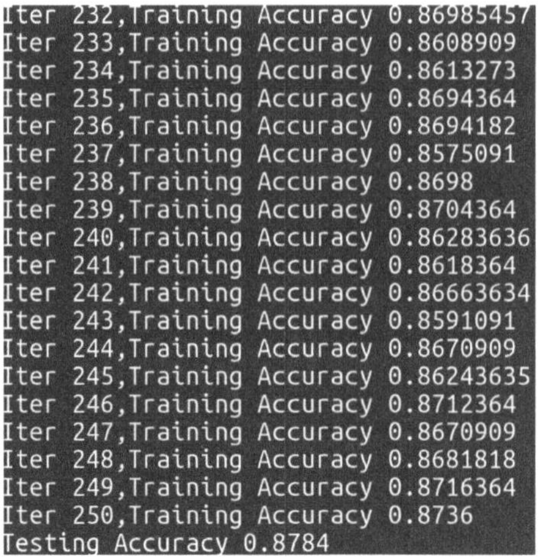

图 9-5 TensorFlow 训练测试结果

kernel 程序代码。

(1) 卷积核的存储格式。

TensorFlow 模型训练后的卷积层的卷积核的存储格式为 $3\times3\times1\times4$,格式如图 9-6(a) 所示。图 9-6(b) 为卷积核 txt 文件中卷积核的存储地址编号,转换为 4 个通道的卷积核后,每个卷积核中的数据在 txt 文件中的地址编号如图 9-6(c) 所示。每个通道的卷积核与图 9-2 中的数据分别做卷积运算。

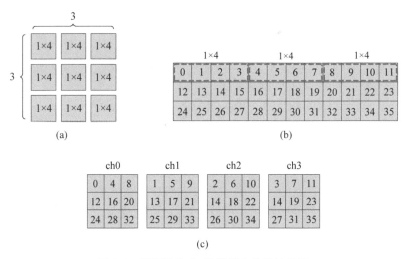

图 9-6 卷积核格式、通道划分及地址编号

(2) 池化结果的存储格式。

池化之后得到的数据格式为 $14\times14\times1\times4$,如图 9-7(a) 所示。图 9-7(b) 表示池化后所有数据(4 个通道)的存储地址编号。为了配合图 9-8 所示的全连接层 1 的权值格式,池化后得到的数据在划分为 4 个通道后,数据存储器地址编号如图 9-7(c) 所示,图中只描述了 ch0 的存储地址,ch1、ch2 和 ch3 的存储地址在 ch0 存储地址的基础上分别+1、+2 和+3。

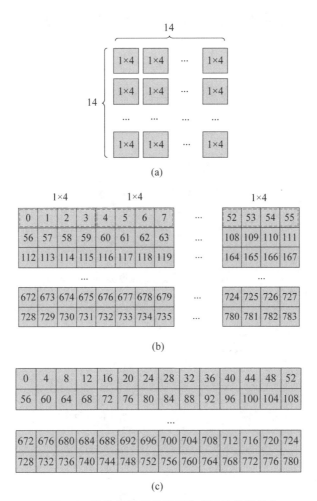

图 9-7 池化层输出的数据格式及地址编号

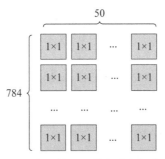

图 9-8 全连接层 1 的权值数据格式

9.3 简易卷积神经网络算法的 OpenCL 实现

9.3.1 NDRange 实现

为了简化设计,在 OpenCL 实现时,将卷积层和池化层用一个 kernel 实现。

1. kernel 程序代码

具体代码如下所示。

```
__kernel void conv1(__global const float * restrict dev_x, //30*30,host 对 28
×28 的数据进行处理,右侧补两列 0,下端补两行 0,30*30
                    __global const float * restrict dev_w1, //3*3*1×4
                    __global const float * restrict dev_b1, //4
                    __global float * restrict dev_y1) //14*14*4
{
  int ch_id=get_global_id(0); //0,1,2,3 通道数量 channel_num
  float result[4][4]={{0.0}};
  int add_index[4]={0,1,30,31};
  float tmp1[4],tmp2[4],tmp3[4]={0.0};
  for (int row=0;row<14;row++)
  {
    for(int col=0;col<14;col++)
    {
      for (int k=0;k<4;k++)              //池化窗为 2×2=4
      {
        result[ch_id][k]=0;
        for(int i=0;i<3;i++)              //行,纵向
        {
          for(int j=0;j<3;j++)            //列,横向
          {
            result[ch_id][k]=result[ch_id][k]+dev_x[row*30*2+col*
            2+30*i+j+add_index[k]]*dev_w1[ch_id+(4*i+12*j)];
          }
        }
        result[ch_id][k]=result[ch_id][k]+dev_b1[ch_id];
        result[ch_id][k]=((result[ch_id][k]>0)? result[ch_id][k]:0);
                                          //Relu 函数
      }

      //pool
      tmp1[ch_id]=(result[ch_id][0]>result[ch_id][1])? result[ch_id][0]:
result[ch_id][1];
      tmp2[ch_id]=(result[ch_id][2]>result[ch_id][3])? result[ch_id][2]:
result[ch_id][3];
```

```
            tmp3[ch_id]=(tmp1[ch_id]>tmp2[ch_id])? tmp1[ch_id]:tmp2[ch_id];

            dev_y1[ch_id+row*14*4+4*col]=tmp3[ch_id];//输出给fc1,格式为14×14×4

        }
      }
    }
}
__kernel void fc1(__global float * restrict dev_y1,      //14×14×4
                  __global float * restrict dev_w2,      //14×14×4×50
                  __global float * restrict dev_b2,      //50
                  __global float * restrict dev_y2)
{
  int ch_id=get_global_id(0);                            //0,1,..,49
  float result[50]={0.0};

  for(int i=0;i<196*4;i++)
    {
       result[ch_id]=result[ch_id]+dev_y1[i]*dev_w2[ch_id+i*50];
    }

    result[ch_id]=result[ch_id]+dev_b2[ch_id];
    result[ch_id]=(result[ch_id]>0)? result[ch_id]:0;
    dev_y2[ch_id]=result[ch_id];

}
//__attribute__((reqd_work_group_size(1,1,1)))
__kernel void fc2(__global float * restrict dev_y2,      //50
                  __global float * restrict dev_w3,      //50*10
                  __global float * restrict dev_b3,      //10
                  __global float * restrict dev_y3)
{

int ch_id=get_global_id(0);                              //0,1,...,9
float result[10]={0.0};

for (int i=0;i<50;i++)                                   //50为fc1通道数量
{
    result[ch_id]=result[ch_id]+dev_y2[i]*dev_w3[ch_id+10*i];
}
    result[ch_id]=result[ch_id]+dev_b3[ch_id];
    dev_y3[ch_id]=result[ch_id];
}
```

2. host 程序代码

具体代码如下所示。

```cpp
#include <assert.h>
#include <stdio.h>
#include <stdlib.h>
#include <math.h>
#include <cstring>
#include "CL/opencl.h"
#include "AOCLUtils/aocl_utils.h"
#include <iostream>
#include <fstream>

#define total_number_image 101
#define image_size   784
#define x_size 28 * 28
#define x1_size 30 * 30
#define w1_size 36
#define b1_size 4
#define y1_size 196 * 4
#define w2_size 196 * 4 * 50
#define b2_size 50
#define y2_size 50
#define w3_size 50 * 10
#define b3_size 10
#define y3_size 10

using namespace aocl_utils;
using namespace std;

//OpenCL runtime configuration
static cl_platform_id platform=NULL;
static cl_device_id device=NULL;
static cl_context context=NULL;
static cl_command_queue queue=NULL;
static cl_kernel mnist_kernel1,mnist_kernel2,mnist_kernel3=NULL;
static cl_program program=NULL;

static cl_mem dev_x,dev_w1,dev_b1,dev_y1,dev_w2,dev_b2,dev_y2,dev_w3,dev_b3,dev_y3=NULL;

//Function prototypes
void ReadFloat(char * filename,  cl_float * data);
```

```cpp
double GetKernelExecutionTime(cl_command_queue cmd, cl_event event, char *
eventname);

const char * kernel_name1="conv1";
const char * kernel_name2="fc1";
const char * kernel_name3="fc2";
const char * source_file="mnist.cl";
const char * aocx_file="mnist";

char input_file_name[5];
char suffix[5]=".txt";
int label_in[1];

cl_int status;

int img_rec_suc=0;

float correct_ratio=0;
float kernel1_execution_time[total_number_image];
float kernel2_execution_time[total_number_image];
float kernel3_execution_time[total_number_image];

int main() {
//Get the OpenCL platform.
    clGetPlatformIDs(1, &platform, NULL);
//Query the available OpenCL devices.
    clGetDeviceIDs(platform, CL_DEVICE_TYPE_ALL, 1,&device,NULL);
//Create the context.
  context=clCreateContext(NULL, 1, &device, NULL, NULL, &status);

//Create the command queue.
  queue=clCreateCommandQueue(context, device, CL_QUEUE_PROFILING_ENABLE,
&status);

//Create the program.
  std::string binary_file=getBoardBinaryFile(aocx_file, device);
  program=createProgramFromBinary(context, binary_file.c_str(), &device, 1);

//Build the program that was just created.
  status=clBuildProgram(program, 0, NULL, "", NULL, NULL);
/*******************************************************************/

double time1=getCurrentTimestamp();
```

```c
/*********************************************************************/

//create the kernel
  mnist_kernel1=clCreateKernel(program, kernel_name1, &status);
  mnist_kernel2=clCreateKernel(program, kernel_name2, &status);
  mnist_kernel3=clCreateKernel(program, kernel_name3, &status);

//allocate and initialize the input vectors

    cl_float * x, * x1, * w1, * b1, * y1, * w2, * b2, * y2, * w3, * b3, * y3;
    x=(cl_float *)alignedMalloc(sizeof(cl_float) * x_size);//
    x1=(cl_float *)alignedMalloc(sizeof(cl_float) * x1_size);//
    w1=(cl_float *)alignedMalloc(sizeof(cl_float) * w1_size);//
    b1=(cl_float *)alignedMalloc(sizeof(cl_float) * b1_size);//
    y1=(cl_float *)alignedMalloc(sizeof(cl_float) * y1_size);//
    w2=(cl_float *)alignedMalloc(sizeof(cl_float) * w2_size);//
    b2=(cl_float *)alignedMalloc(sizeof(cl_float) * b2_size);//
    y2=(cl_float *)alignedMalloc(sizeof(cl_float) * y2_size);//
    w3=(cl_float *)alignedMalloc(sizeof(cl_float) * w3_size);//
    b3=(cl_float *)alignedMalloc(sizeof(cl_float) * b3_size);//
    y3=(cl_float *)alignedMalloc(sizeof(cl_float) * y3_size);//

//create the input buffer

    cl_mem dev_x1,dev_w1,dev_b1,dev_y1,dev_w2,dev_b2,dev_y2,dev_w3,dev_b3,
dev_y3;

dev_x1=clCreateBuffer(context, CL_MEM_READ_WRITE, sizeof(cl_float) * x1_
size, NULL, &status);
    dev_w1=clCreateBuffer(context, CL_MEM_READ_WRITE, sizeof(cl_float) * w1_
size, NULL, &status);
    dev_b1=clCreateBuffer(context, CL_MEM_READ_WRITE, sizeof(cl_float) * b1_
size, NULL, &status);
    dev_y1=clCreateBuffer(context, CL_MEM_READ_WRITE, sizeof(cl_float) * y1_
size, NULL, &status);
    dev_w2=clCreateBuffer(context, CL_MEM_READ_WRITE, sizeof(cl_float) * w2_
size, NULL, &status);
    dev_b2=clCreateBuffer(context, CL_MEM_READ_WRITE, sizeof(cl_float) * b2_
size, NULL, &status);
    dev_y2=clCreateBuffer(context, CL_MEM_READ_WRITE, sizeof(cl_float) * y2_
size, NULL, &status);

dev_w3=clCreateBuffer(context, CL_MEM_READ_WRITE, sizeof(cl_float) * w3_
size, NULL, &status);
```

```
    dev_b3=clCreateBuffer(context, CL_MEM_READ_WRITE, sizeof(cl_float) * b3_
size, NULL, &status);
    dev_y3=clCreateBuffer(context, CL_MEM_READ_WRITE, sizeof(cl_float) * y3_
size, NULL, &status);

//load data from text file
    ReadFloat("w_conv1.txt",w1);
    ReadFloat("b_conv1.txt",b1);
    ReadFloat("w_fc1.txt", w2);
    ReadFloat("b_fc1.txt", b2);
    ReadFloat("w_fc2.txt", w3);
    ReadFloat("b_fc2.txt", b3);

//Write constant buffer
    status=clEnqueueWriteBuffer(queue, dev_w1, CL_TRUE, 0, sizeof(cl_float) *
w1_size, w1, 0, NULL, NULL);
    status=clEnqueueWriteBuffer(queue, dev_b1, CL_TRUE, 0, sizeof(cl_float) *
b1_size, b1, 0, NULL, NULL);

    status=clEnqueueWriteBuffer(queue, dev_w2, CL_TRUE, 0, sizeof(cl_float) *
w2_size, w2, 0, NULL, NULL);
    status=clEnqueueWriteBuffer(queue, dev_b2, CL_TRUE, 0, sizeof(cl_float) *
b2_size, b2, 0, NULL, NULL);

    status=clEnqueueWriteBuffer(queue, dev_w3, CL_TRUE, 0, sizeof(cl_float) *
w3_size, w3, 0, NULL, NULL);
    status=clEnqueueWriteBuffer(queue, dev_b3, CL_TRUE, 0, sizeof(cl_float) *
b3_size, b3, 0, NULL, NULL);

//read the input image file
    for (int img_index=0;img_index<int(total_number_image);img_index++)
    {
        sprintf(input_file_name,"%d",img_index);

        char input_file_path[45]="./mnist_txt/mnist_img_txt/img_";
        char input_lab_path[45]="./mnist_txt/mnist_lab_txt/img_lab_";

    printf("##################################################\n");
    printf("    input_filename=\033[7mimg_%s.txt\033[0m\n",input_file_name);

//reading data from text file
    ReadFloat(strcat(strcat(input_file_path,input_file_name),suffix),
x);//
```

```c
//pading for input data

    for(int row=0;row<30;row++)
        {
            for(int col=0;col<30;col++)
            {
                x1[30 * row+col]=0.0;
            }
        }

    for(int row=0;row<28;row++)
        {
            for(int col=0;col<28;col++)
            {
                x1[28 * row+col+2 * row]=x[28 * row+col];
            }
        }

/////////////////////////////////////////////////////////////////////
    printf("**********************************************\n");
    printf("\033[7m \033[40;31m ****read image sucessful********\033[0m\n");
    printf("**********************************************\n");
    /////////////////////////////////////////////////////

    printf("**********************************************\n");
    printf("*     input_lab_filename=\033[7mimg_lab_%s.txt \033[0m\n",
input_file_name);
    printf("**********************************************\n");

//read the label of input image file

        FILE * fp1;
        fp1=fopen(strcat(strcat(input_lab_path,input_file_name),suffix),"r");

    fscanf(fp1,"%i",label_in);
    fclose(fp1);
    printf("**********************************************\n");
    printf("*      \033[7m label of input image=%i    \033[0m\n",label_in[0]);
    printf("**********************************************\n");

//Write input image buffer
    status=clEnqueueWriteBuffer(queue, dev_x1, CL_TRUE, 0, sizeof(cl_float)
* x1_size, x1, 0, NULL, NULL);
```

```c
//set the arguments
  status=clSetKernelArg(mnist_kernel1,0, sizeof(cl_mem), (void*)&dev_x1);
  status=clSetKernelArg(mnist_kernel1,1, sizeof(cl_mem), (void*)&dev_w1);
  status=clSetKernelArg(mnist_kernel1,2, sizeof(cl_mem), (void*)&dev_b1);
  status=clSetKernelArg(mnist_kernel1,3, sizeof(cl_mem), (void*)&dev_y1);

  status=clSetKernelArg(mnist_kernel2,0, sizeof(cl_mem), (void*)&dev_y1);
  status=clSetKernelArg(mnist_kernel2,1, sizeof(cl_mem), (void*)&dev_w2);
  status=clSetKernelArg(mnist_kernel2,2, sizeof(cl_mem), (void*)&dev_b2);
  status=clSetKernelArg(mnist_kernel2,3, sizeof(cl_mem), (void*)&dev_y2);

  status=clSetKernelArg(mnist_kernel3,0, sizeof(cl_mem), (void*)&dev_y2);
  status=clSetKernelArg(mnist_kernel3,1, sizeof(cl_mem), (void*)&dev_w3);
  status=clSetKernelArg(mnist_kernel3,2, sizeof(cl_mem), (void*)&dev_b3);
  status=clSetKernelArg(mnist_kernel3,3, sizeof(cl_mem), (void*)&dev_y3);

//launch kernel1
  static const size_t GSize1[]={4};   //mnist_simple 的 global size
  static const size_t WSize1[]={1};   //mnist_simple 的 local size
  cl_event event_kernel1;
  char dim=1;

  status=clEnqueueNDRangeKernel(queue, mnist_kernel1, dim,0, GSize1, WSize1, 0, NULL, &event_kernel1);

/*************count runtime of kernel**************************/
  double time_sum1=0;
  time_sum1=GetKernelExecutionTime(queue, event_kernel1,"event_cluster");//
  kernel1_execution_time[img_index]=time_sum1/1000000;
/***************************************************************/

//read the output of kernel1
    status=clEnqueueReadBuffer(queue, dev_y1, CL_TRUE, 0, sizeof(cl_float) * y1_size, y1, 0, NULL, NULL);

//launch kernel2
  static const size_t GSize2[]={50};  //mnist_simple 的 global size
  static const size_t WSize2[]={1};   //mnist_simple 的 local size
  cl_event event_kernel2;

   status = clEnqueueNDRangeKernel (queue, mnist_kernel2, dim, 0, GSize2, WSize2, 0, NULL, &event_kernel2);
```

```c
/*************count runtime of kernel*************************/
  double time_sum2=0;
  time_sum2=GetKernelExecutionTime(queue, event_kernel2,"event_cluster");//
  kernel2_execution_time[img_index]=time_sum2/1000000;


/***************************************************************/

//read the output of kernel2
    status=clEnqueueReadBuffer(queue, dev_y2, CL_TRUE, 0, sizeof(cl_float) * y2_size, y2 , 0, NULL, NULL);

//launch kernel3
  static const size_t GSize3[]={10};   //mnist_simple 的 global size
  static const size_t WSize3[]={1};    //mnist_simple 的 local size
  cl_event event_kernel3;

   status = clEnqueueNDRangeKernel (queue, mnist_kernel3, dim, 0, GSize3, WSize3, 0, NULL, &event_kernel3);

/*************count runtime of kernel*************************/
  double time_sum3=0;
  time_sum3=GetKernelExecutionTime(queue, event_kernel3,"event_cluster");//
  kernel3_execution_time[img_index]=time_sum3/1000000;


/***************************************************************/

//read the output of kernel3
    status=clEnqueueReadBuffer(queue, dev_y3, CL_TRUE, 0, sizeof(cl_float) * y3_size, y3 , 0, NULL, NULL);

//display result
        for(int j=0;j<y3_size;j++)
            {
            printf("j=%i,",j);
            printf("number=%.16f\n",y3[j]);
            }
//display recognaize label
        cl_float tmp=0;
        int lab;

        for(int j=0;j<10;j++)
            {
                if (y3[j]>tmp)
```

```
                {
                tmp=y3[j];
                lab=j;
                }
        }
printf("*********************************************\n");
printf(" *      \033[7m label_recognized=%i \033[0m\n",lab);
printf("*********************************************\n");
   if(lab==label_in[0]) {
       img_rec_suc=img_rec_suc+1;
   }
}

/**********************************************************************/
double time2=getCurrentTimestamp();
/**********************************************************************/
  correct_ratio=float(img_rec_suc)/int(total_number_image);
  printf("*********************************************\n");
  printf(" *      img_input_num=%i\n",total_number_image);
  printf(" *      img_rec_suc_num=\033[7m%i\033[0m       \n",img_rec_suc);
  printf(" *      correct ratio=\033[7m%f\033[0m        \n",correct_ratio);
  printf("*********************************************\n");

/**********************************************************************/
//print runtime of kernel

printf("kernel_execution_time is:\n");
for(int i=0;i<total_number_image;i++)
printf("%f\n",kernel1_execution_time[i]+kernel2_execution_time[i]+kernel3_execution_time[i]);

printf("total_time is:%f ms\n",(time2-time1) * 1e3);

///////////////////////////
  clFlush(queue);
  clFinish(queue);
  //device side
  clReleaseMemObject(dev_x);
  clReleaseMemObject(dev_w1);
  clReleaseMemObject(dev_b1);
  clReleaseMemObject(dev_y1);
  clReleaseMemObject(dev_w2);
  clReleaseMemObject(dev_b2);
```

```c
    clReleaseMemObject(dev_y2);
    clReleaseMemObject(dev_w3);
    clReleaseMemObject(dev_b3);
    clReleaseMemObject(dev_y3);
    clReleaseKernel(mnist_kernel1);
    clReleaseKernel(mnist_kernel2);
    clReleaseKernel(mnist_kernel3);

    clReleaseProgram(program);
    clReleaseCommandQueue(queue);
    clReleaseContext(context);
//hose side
    free(x);
    free(w1);
    free(b1);
    free(y1);
    free(w2);
    free(b2);
    free(y2);
    free(w3);
    free(b3);
    free(y3);

///////////////////////////
    return 0;
}

//******************************* Added by Page*************************//
void ReadFloat(char * filename, cl_float * data)
{
    FILE * fp1;
    fp1=fopen(filename,"r+");
    int j=0;
    while(fscanf(fp1,"%f",&data[j++])!=-1);
    fclose(fp1);
}

void cleanup()
{
}
double GetKernelExecutionTime(cl_command_queue cmd, cl_event event, char * eventname)
{
```

```
    cl_ulong start,end;
    clFinish(cmd);
    clGetEventProfilingInfo(event,CL_PROFILING_COMMAND_START,sizeof(cl_
ulong),&start,NULL);
    clGetEventProfilingInfo(event,CL_PROFILING_COMMAND_END,sizeof(cl_
ulong),&end,NULL);
    double runtime=(double)(end-start);
    return runtime;
}
```

3. 运行结果

运行结果如图 9-9 所示。

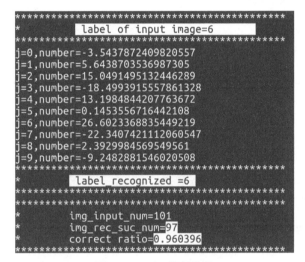

图 9-9　FPGA 运行结果

4. 资源使用

资源使用情况如图 9-10 所示。

```
+-------------------------------------------------------+
; Estimated Resource Usage Summary                      ;
+----------------------------------+--------------------+
; Resource                         ; + Usage            ;
+----------------------------------+--------------------+
; Logic utilization                ;    65%             ;
; ALUTs                            ;    44%             ;
; Dedicated logic registers        ;    26%             ;
; Memory blocks                    ;    87%             ;
; DSP blocks                       ;    10%             ;
+----------------------------------+--------------------+
```

图 9-10　资源使用情况

5. kernel 运行时间

kernel 运行时间如图 9-11 所示。

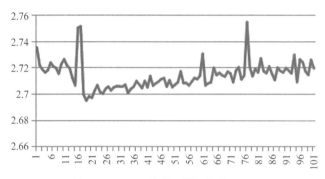

图 9-11 kernel 运行时间（单位：ms）

9.3.2 single work item 实现

1. kernel 程序代码

具体代码如下所示。

```
__kernel void conv1(__global const float * restrict dev_x,   //30 * 30,host 对
28×28的数据进行处理,右侧补两列 0,下端补两行 0,30 * 30
                    __global const float * restrict dev_w1,  //3 * 3 * 1×4
                    __global const float * restrict dev_b1,  //4
                    __global float * restrict dev_y1)        //14 * 14 * 4
{

  float result[4][4]={{0.0}};
  int add_index[4]={0,1,30,31};
  float tmp1[4],tmp2[4],tmp3[4]={0.0};

for (int ch_id=0;ch_id<4;ch_id++)           //0,1,2,3 通道数量
{
  for (int row=0;row<14;row++)
    {
      for(int col=0;col<14;col++)
      {

        for (int k=0;k<4;k++)
        {
          result[ch_id][k]=0;
          for(int i=0;i<3;i++)              //行,纵向
          {
            for(int j=0;j<3;j++)            //列,横向
            {
              result[ch_id][k]=result[ch_id][k]+dev_x[row * 30 * 2+col * 2+30 *
i+j+add_index[k]] * dev_w1[ch_id+(4 * i+12 * j)];
```

```
                }
            }
            result[ch_id][k]=result[ch_id][k]+dev_b1[ch_id];
            result[ch_id][k]=((result[ch_id][k]>0)? result[ch_id][k]:0);
                                                                    //Relu函数
        }

    //pool

        tmp1[ch_id]=(result[ch_id][0]>result[ch_id][1])? result[ch_id][0]:result[ch_id][1];
        tmp2[ch_id]=(result[ch_id][2]>result[ch_id][3])? result[ch_id][2]:result[ch_id][3];
        tmp3[ch_id]=(tmp1[ch_id]>tmp2[ch_id])? tmp1[ch_id]:tmp2[ch_id];

        dev_y1[ch_id+row*14*4+4*col]=tmp3[ch_id]; //输出给 fc1,格式为 14×14×4
        }
      }
    }
}

__kernel void fc1(__global float * restrict dev_y1,         //14×14×4
                  __global float * restrict dev_w2,         //14×14×4×50
                  __global float * restrict dev_b2,         //50
                  __global float * restrict dev_y2)
{
    float result[50]={0.0};
    for(int ch_id=0;ch_id<50;ch_id++)//0,1,2,3,...,49
    {
        for(int i=0;i<196*4;i++)//14×14×4 为 conv1 的格式
        {
            result[ch_id]=result[ch_id]+dev_y1[i]*dev_w2[ch_id+i*50];
        }

        result[ch_id]=result[ch_id]+dev_b2[ch_id];
        result[ch_id]=(result[ch_id]>0)? result[ch_id]:0;
        dev_y2[ch_id]=result[ch_id];

    }
}
```

```
__kernel void fc2(__global float * restrict dev_y2,//50
                  __global float * restrict dev_w3,//50*10
                  __global float * restrict dev_b3, //10
                  __global float * restrict dev_y3)
{

float result[10]={0.0};

for (int ch_id=0;ch_id<10;ch_id++)
{
for (int i=0;i<50;i++)
    {
      result[ch_id]=result[ch_id]+dev_y2[i]*dev_w3[ch_id+10*i];
    }
    result[ch_id]=result[ch_id]+dev_b3[ch_id];
    dev_y3[ch_id]=result[ch_id];
  }
}
```

2. host 程序代码

具体代码如下所示。

```
#include <assert.h>
#include <stdio.h>
#include <stdlib.h>
#include <math.h>
#include <cstring>
#include "CL/opencl.h"
#include "AOCLUtils/aocl_utils.h"
#include <iostream>
#include <fstream>

#define total_number_image 101
#define image_size784
#define x_size 28*28
#define x1_size 30*30
#define w1_size 36
#define b1_size 4
#define y1_size 196*4
#define w2_size 196*4*50
#define b2_size 50
#define y2_size 50
```

```cpp
#define w3_size 50*10
#define b3_size 10
#define y3_size 10

using namespace aocl_utils;
using namespace std;

//OpenCL runtime configuration
static cl_platform_id platform=NULL;
static cl_device_id device=NULL;
static cl_context context=NULL;
static cl_command_queue queue=NULL;
static cl_kernel mnist_kernel1,mnist_kernel2,mnist_kernel3=NULL;
static cl_program program=NULL;
static cl_mem dev_x,dev_w1,dev_b1,dev_y1,dev_w2,dev_b2,dev_y2,dev_w3,dev_b3,dev_y3=NULL;

//Function prototypes

void ReadFloat(char * filename,  cl_float * data);
double GetKernelExecutionTime(cl_command_queue cmd, cl_event event, char * eventname);

//
const char * kernel_name1="conv1";
const char * kernel_name2="fc1";
const char * kernel_name3="fc2";
const char * source_file="mnist.cl";
const char * aocx_file="mnist";

char input_file_name[5];
char suffix[5]=".txt";
int label_in[1];

cl_int status;

int img_rec_suc=0;

float correct_ratio=0;
float kernel1_execution_time[total_number_image];
float kernel2_execution_time[total_number_image];
float kernel3_execution_time[total_number_image];
```

```c
int main() {
//Get the OpenCL platform.
    clGetPlatformIDs(1, &platform, NULL);
//Query the available OpenCL devices.
    clGetDeviceIDs(platform, CL_DEVICE_TYPE_ALL, 1,&device,NULL);
//Create the context.
   context=clCreateContext(NULL, 1, &device, NULL, NULL, &status);

//Create the command queue.
   queue=clCreateCommandQueue(context, device, CL_QUEUE_PROFILING_ENABLE,
&status);

//Create the program.
   std::string binary_file=getBoardBinaryFile(aocx_file, device);
   program=createProgramFromBinary(context, binary_file.c_str(), &device, 1);

//Build the program that was just created.
   status=clBuildProgram(program, 0, NULL, "", NULL, NULL);
/**********************************************************************/
double time1=getCurrentTimestamp();
/**********************************************************************/
//create the kernel
   mnist_kernel1=clCreateKernel(program, kernel_name1, &status);
   mnist_kernel2=clCreateKernel(program, kernel_name2, &status);
   mnist_kernel3=clCreateKernel(program, kernel_name3, &status);

//allocate and initialize the input vectors

    cl_float * x, * x1, * w1, * b1, * y1, * w2, * b2, * y2, * w3, * b3, * y3;
    x=(cl_float * )alignedMalloc(sizeof(cl_float) * x_size);//
    x1=(cl_float * )alignedMalloc(sizeof(cl_float) * x1_size);//
    w1=(cl_float * )alignedMalloc(sizeof(cl_float) * w1_size);//
    b1=(cl_float * )alignedMalloc(sizeof(cl_float) * b1_size);//
    y1=(cl_float * )alignedMalloc(sizeof(cl_float) * y1_size);//
    w2=(cl_float * )alignedMalloc(sizeof(cl_float) * w2_size);//
    b2=(cl_float * )alignedMalloc(sizeof(cl_float) * b2_size);//
    y2=(cl_float * )alignedMalloc(sizeof(cl_float) * y2_size);//
    w3=(cl_float * )alignedMalloc(sizeof(cl_float) * w3_size);//
    b3=(cl_float * )alignedMalloc(sizeof(cl_float) * b3_size);//
    y3=(cl_float * )alignedMalloc(sizeof(cl_float) * y3_size);//

//create the input buffer
    cl_mem dev_x1,dev_w1,dev_b1,dev_y1,dev_w2,dev_b2,dev_y2,dev_w3,dev_b3,
dev_y3;
```

```
    dev_x1=clCreateBuffer(context, CL_MEM_READ_WRITE, sizeof(cl_float) * x1_
size, NULL, &status);
    dev_w1=clCreateBuffer(context, CL_MEM_READ_WRITE, sizeof(cl_float) * w1_
size, NULL, &status);
    dev_b1=clCreateBuffer(context, CL_MEM_READ_WRITE, sizeof(cl_float) * b1_
size, NULL, &status);
    dev_y1=clCreateBuffer(context, CL_MEM_READ_WRITE, sizeof(cl_float) * y1_
size, NULL, &status);
     dev_w2=clCreateBuffer(context, CL_MEM_READ_WRITE, sizeof(cl_float) * w2
_size, NULL, &status);
    dev_b2=clCreateBuffer(context, CL_MEM_READ_WRITE, sizeof(cl_float) * b2_
size, NULL, &status);
    dev_y2=clCreateBuffer(context, CL_MEM_READ_WRITE, sizeof(cl_float) * y2_
size, NULL, &status);
     dev_w3=clCreateBuffer(context, CL_MEM_READ_WRITE, sizeof(cl_float) * w3
_size, NULL, &status);
    dev_b3=clCreateBuffer(context, CL_MEM_READ_WRITE, sizeof(cl_float) * b3_
size, NULL, &status);
    dev_y3=clCreateBuffer(context, CL_MEM_READ_WRITE, sizeof(cl_float) * y3_
size, NULL, &status);

//load data from text file
    ReadFloat("w_conv1.txt",w1);
    ReadFloat("b_conv1.txt",b1);
    ReadFloat("w_fc1.txt", w2);
    ReadFloat("b_fc1.txt", b2);
    ReadFloat("w_fc2.txt", w3);
    ReadFloat("b_fc2.txt", b3);
//Write constant buffer
    status=clEnqueueWriteBuffer(queue, dev_w1, CL_TRUE, 0, sizeof(cl_float) *
w1_size, w1, 0, NULL, NULL);
    status=clEnqueueWriteBuffer(queue, dev_b1, CL_TRUE, 0, sizeof(cl_float) *
b1_size, b1, 0, NULL, NULL);
    status=clEnqueueWriteBuffer(queue, dev_w2, CL_TRUE, 0, sizeof(cl_float) *
w2_size, w2, 0, NULL, NULL);
    status=clEnqueueWriteBuffer(queue, dev_b2, CL_TRUE, 0, sizeof(cl_float) *
b2_size, b2, 0, NULL, NULL);
    status=clEnqueueWriteBuffer(queue, dev_w3, CL_TRUE, 0, sizeof(cl_float) *
w3_size, w3, 0, NULL, NULL);
    status=clEnqueueWriteBuffer(queue, dev_b3, CL_TRUE, 0, sizeof(cl_float) *
b3_size, b3, 0, NULL, NULL);

//read the input image file
```

```c
    for (int img_index=0;img_index<int(total_number_image);img_index++)
    {
        sprintf(input_file_name,"%d",img_index);
        char input_file_path[45]="./mnist_txt/mnist_img_txt/img_";
        char input_lab_path[45]="./mnist_txt/mnist_lab_txt/img_lab_";

        printf("##################################################\n");
        printf("input_filename=\033[7mimg_%s.txt\033[0m\n",input_file_name);

//reading data from text file
        ReadFloat(strcat(strcat(input_file_path,input_file_name),suffix),x);//
//pading for input data

    for(int row=0;row<30;row++)
    {
        for(int col=0;col<30;col++)
        {
            x1[30 * row+col]=0.0;
        }
    }

    for(int row=0;row<28;row++)
    {
        for(int col=0;col<28;col++)
        {
            //x1[28 * row+col+2 * row]=x[28 * row+col];
            x1[30 * (row+1)+col+1]=x[28 * row+col];
        }
    }

//////////////////////////////////////////////////////////////////////////
        printf("**********************************************\n");
        printf("\033[7m \033[40;31m ****read image sucessful********\033[0m\n");
        printf("**********************************************\n");
//////////////////////////////////////////////////
        printf("**********************************************\n");
        printf("*        input_lab_filename=\033[7mimg_lab_%s.txt \033[0m\n",input_file_name);
        printf("**********************************************\n");

//read the label of input image file
        FILE * fp1;
        fp1=fopen(strcat(strcat(input_lab_path,input_file_name),suffix),"r");//
```

```c
    fscanf(fp1,"%i",label_in);
    fclose(fp1);
    printf("**************************************************\n");
    printf("*      \033[7m label of input image=%i     \033[0m\n",label_in[0]);
    printf("**************************************************\n");

//Write input image buffer
    status=clEnqueueWriteBuffer(queue, dev_x1, CL_TRUE, 0, sizeof(cl_float) *
x1_size, x1, 0, NULL, NULL);

//set the arguments
    status=clSetKernelArg(mnist_kernel1,0, sizeof(cl_mem), (void*)&dev_x1);
    status=clSetKernelArg(mnist_kernel1,1, sizeof(cl_mem), (void*)&dev_w1);
    status=clSetKernelArg(mnist_kernel1,2, sizeof(cl_mem), (void*)&dev_b1);
    status=clSetKernelArg(mnist_kernel1,3, sizeof(cl_mem), (void*)&dev_y1);

    status=clSetKernelArg(mnist_kernel2,0, sizeof(cl_mem), (void*)&dev_y1);
    status=clSetKernelArg(mnist_kernel2,1, sizeof(cl_mem), (void*)&dev_w2);
    status=clSetKernelArg(mnist_kernel2,2, sizeof(cl_mem), (void*)&dev_b2);
    status=clSetKernelArg(mnist_kernel2,3, sizeof(cl_mem), (void*)&dev_y2);

    status=clSetKernelArg(mnist_kernel3,0, sizeof(cl_mem), (void*)&dev_y2);
    status=clSetKernelArg(mnist_kernel3,1, sizeof(cl_mem), (void*)&dev_w3);
    status=clSetKernelArg(mnist_kernel3,2, sizeof(cl_mem), (void*)&dev_b3);
    status=clSetKernelArg(mnist_kernel3,3, sizeof(cl_mem), (void*)&dev_y3);

//launch kernel1
    static const size_t GSize1[]={1};   //mnist_simple的 global size
    static const size_t WSize1[]={1};   //mnist_simple的 local size
    cl_event event_kernel1;
    char dim=1;

    status=clEnqueueNDRangeKernel(queue, mnist_kernel1, dim,0, GSize1, WSize1, 0,
NULL, &event_kernel1);

/*************count runtime of kernel*************************/
    double time_sum1=0;
    time_sum1=GetKernelExecutionTime(queue, event_kernel1,"event_cluster");//
    kernel1_execution_time[img_index]=time_sum1/1000000;
/**************************************************************/

//read the output of kernel1
    status=clEnqueueReadBuffer(queue, dev_y1, CL_TRUE, 0, sizeof(cl_float) *
y1_size, y1 , 0, NULL, NULL);
```

```c
//launch kernel2
  static const size_t GSize2[]={1};  //mnist_simple的global size
  static const size_t WSize2[]={1}; //mnist_simple的local size
  cl_event event_kernel2;

   status = clEnqueueNDRangeKernel (queue, mnist_kernel2, dim, 0, GSize2, WSize2, 0, NULL, &event_kernel2);

/*************count runtime of kernel*************************/
  double time_sum2=0;
   time_sum2 =  GetKernelExecutionTime (queue, event_kernel2," event_cluster");//
   kernel2_execution_time[img_index]=time_sum2/1000000;
/***************************************************************/

//read the output of kernel2
    status=clEnqueueReadBuffer(queue, dev_y2, CL_TRUE, 0, sizeof(cl_float) * y2_size, y2 , 0, NULL, NULL);

//launch kernel3
  static const size_t GSize3[]={1};  //mnist_simple的global size
  static const size_t WSize3[]={1};  //mnist_simple的local size
  cl_event event_kernel3;

   status = clEnqueueNDRangeKernel (queue, mnist_kernel3, dim, 0, GSize3, WSize3, 0, NULL, &event_kernel3);

/*************count runtime of kernel*************************/
  double time_sum3=0;
  time_sum3=GetKernelExecutionTime(queue, event_kernel3,"event_cluster");//
   kernel3_execution_time[img_index]=time_sum3/1000000;

/***************************************************************/

//read the output of kernel3
    status=clEnqueueReadBuffer(queue, dev_y3, CL_TRUE, 0, sizeof(cl_float) * y3_size, y3 , 0, NULL, NULL);
//display result
        for(int j=0;j<y3_size;j++)
            {
              printf("j=%i,",j);
              printf("number=%.16f\n",y3[j]);
            }
```

```
//display recognaize label
    cl_float tmp=0;
    int lab;

    for(int j=0;j<10;j++)
        {
           if(y3[j]>tmp)
             {
                tmp=y3[j];
                lab=j;
             }


        }
printf("**********************************************\n");
printf("*    \033[7m label_recognized=%i \033[0m\n",lab);
printf("**********************************************\n");
  if (lab==label_in[0]) {
      img_rec_suc=img_rec_suc+1;
  }



}

/******************************************************************/

double time2=getCurrentTimestamp();
/******************************************************************/
  correct_ratio=float(img_rec_suc)/int(total_number_image);
  printf("**********************************************\n");
  printf("*    img_input_num=%i\n",total_number_image);
  printf("*    img_rec_suc_num=\033[7m%i\033[0m            \n",img_rec_suc);
  printf("*    correct ratio=\033[7m%f\033[0m              \n",correct_ratio);
  printf("**********************************************\n");

/******************************************************************/
//print runtime of kernel

printf("kernel_execution_time is:\n");
for(int i=0;i<total_number_image;i++)
printf("%f\n",kernel1_execution_time[i]+kernel2_execution_time[i]+kernel3_execution_time[i]);
printf("total_time is:%f ms\n",(time2-time1) * 1e3);

///////////////////////////
   clFlush(queue);
```

```
  clFinish(queue);
//device side
  clReleaseMemObject(dev_x);
  clReleaseMemObject(dev_w1);
  clReleaseMemObject(dev_b1);
  clReleaseMemObject(dev_y1);
  clReleaseMemObject(dev_w2);
  clReleaseMemObject(dev_b2);
  clReleaseMemObject(dev_y2);
  clReleaseMemObject(dev_w3);
  clReleaseMemObject(dev_b3);
  clReleaseMemObject(dev_y3);
  clReleaseKernel(mnist_kernel1);
  clReleaseKernel(mnist_kernel2);
  clReleaseKernel(mnist_kernel3);

  clReleaseProgram(program);
  clReleaseCommandQueue(queue);
  clReleaseContext(context);
//hose side
  free(x);
  free(w1);
  free(b1);
  free(y1);
  free(w2);
  free(b2);
  free(y2);
  free(w3);
  free(b3);
  free(y3);
/////////////////////////////
  return 0;
}
//****************************** Added by Page *************************//
void ReadFloat(char *filename, cl_float *data)
{
    FILE *fp1;                    //
    fp1=fopen(filename,"r+");     //
    int j=0;
    while(fscanf(fp1,"%f",&data[j++])!=-1);//
    fclose(fp1);                  //
}
void cleanup()
```

```
{
}
double GetKernelExecutionTime(cl_command_queue cmd, cl_event event, char *
eventname)
{
    cl_ulong start,end;
    clFinish(cmd);
    clGetEventProfilingInfo(event,CL_PROFILING_COMMAND_START,sizeof(cl_
ulong),&start,NULL);
     clGetEventProfilingInfo(event,CL_PROFILING_COMMAND_END,sizeof(cl_
ulong),&end,NULL);
    double runtime=(double)(end-start);
    return runtime;
}
```

3. 运行结果

运行结果如图 9-12 所示。

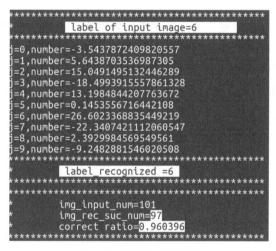

图 9-12　FPGA 运行结果

4. 资源使用

资源使用情况如图 9-13 所示。

```
+------------------------------------------------+
; Estimated Resource Usage Summary               ;
+-----------------------------+------------------+
; Resource                    + Usage            ;
+-----------------------------+------------------+
; Logic utilization           ;    74%           ;
; ALUTs                       ;    47%           ;
; Dedicated logic registers   ;    32%           ;
; Memory blocks               ;    77%           ;
; DSP blocks                  ;    10%           ;
+-----------------------------+------------------+
```

图 9-13　资源使用情况

5. kernel 运行时间

kernel 运行时间如图 9-14 所示。

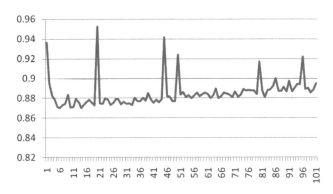

图 9-14　kernel 运行时间（单位：ms）

习题 9

9.1　试分析如何根据卷积步长及 padding 模式确定输入图像的边界处理。

9.2　试分析图 9-3 中卷积运算顺序和格式的依据是什么。

9.3　TensorFlow 模型中卷积核的存储格式有何特点？

9.4　TensorFlow 模型中池化层输出的数据格式有何特点？

9.5　卷积和池化用一个 kernel 实现的好处是什么？

第 10 章

上 机 实 验

本章根据课程内容给出 13 个上机实验,帮助读者通过上机实验进一步理解和掌握课程内容。

注意:上机实验正常运行的前提是相关软件的运行环境、硬件平台已经搭建成功。TensorFlow 运行环境的搭建详见 2.4 节,OpenCL 的运行环境搭建详见第 5 章。

实验 1　TensorFlow 基础命令

1. 实验目的

(1) 熟悉在 Ubuntu 系统下运行 TensorFlow 的基本流程。

(2) 熟悉 TensorFlow 的两步编程模式。

(3) 掌握 TensorFlow 的基本操作语法。

2. 实验内容

运行 2.4.3 节中的计算图例程运行实例并进行代码分析,到达实验目的要求。

实验 2　TensorFlow 实现简易神经网络模型的训练与测试

1. 实验目的

(1) 熟悉神经网络训练与测试的基本概念。

(2) 熟悉 MNIST 手写字体识别的基本原理。

(3) 熟悉基于 TensorFlow 实现神经网络的基本流程。

(4) 进一步掌握 TensorFlow 的基本操作语法。

2. 实验内容

运行 3.2.3 节的实验步骤,阅读 3.2.3 节的代码解读。

实验 3　TensorFlow 实现卷积神经网络模型的训练与测试

1. 实验目的

（1）了解卷积神经网络的基本原理。

（2）进一步了解 TensorFlow 的基本操作语法。

2. 实验内容

运行 3.2.4 节的实验步骤。

实验 4　TensorFlow 实现 MNIST 数据集转换

1. 实验目的

（1）掌握数据集格式转换的基本方法。

（2）进一步了解 TensorFlow 的基本操作语法。

2. 实验内容

（1）将 MNIST 数据集转换为以 txt 文件保存的数据。运行 3.3.1 节的实验步骤。

（2）将 MNIST 数据集转换为以 bmp 文件保存的图片。运行 3.3.2 节的实验步骤。

（3）将 bmp 转换为 tfrecords 格式。运行 3.3.3 节的实验步骤。

实验 5　读取 tfrecords 格式数据并实现 MNIST 手写字体识别

1. 实验目的

（1）掌握读取 tfrecords 格式数据的基本方法。

（2）进一步了解 TensorFlow 的基本操作语法。

2. 实验内容

运行 3.4 节的实验步骤，分析比较其与实验 2、实验 3 代码的异同，对比训练结果。

实验 6　DE10_nano 开发板运行 OpenCL 程序

1. 实验目的

（1）测试 minicom 端口。

（2）掌握 DE10_nano 开发板运行 OpenCL 程序的基本步骤。

2. 实验内容

运行 5.10 节的实验步骤。

实验 7　DE10_nano 与 PC 数据交换

1. 实验目的

掌握 DE10_nano 开发板与 PC 交换数据的基本步骤。

2. 实验内容

运行 5.11 节的实验步骤。

实验 8　OpenCL 程序编译

1. 实验目的

(1) 掌握编译 kernel 程序的基本步骤。
(2) 掌握编译 host 程序的基本步骤。

2. 实验内容

运行 5.6 节和 5.7 节的实验步骤(注：编译 kernel 程序代码需要较长的时间,该实验内容可根据实际情况选做)。

实验 9　编写一个 OpenCL 程序

1. 实验目的

(1) 熟悉 OpenCL 的基本概念。
(2) 掌握 kernel 程序的编写规则及 host 程序的基本流程。

2. 实验内容

结合实验 6、实验 7、实验 8 运行 4.5 节的 OpenCL 代码。

实验 10　单层神经网络算法模型的 FPGA 实现流程

1. 实验目的

(1) 掌握基于 OpenCL 设计神经网络的基本流程。
(2) 熟悉查看 report.html 文件获取设计信息的方法。

2. 实验内容

利用第 6 章的知识,根据课程内容复现图 6-11 所示的 FPGA 运行结果。

实验 11　单层神经网络算法的 kernel 程序的不同实现方式

1. 实验目的

(1) 分析比较 kernel 编程方式、变量存储方式、数据类型对 FPGA 实现的资源使用、神经网络算法运行时间、FPGA 硬件运行效果的影响。

（2）分析比较 ARM 软件实现和 FPGA 硬件实现对 FPGA 实现的资源使用、神经网络算法运行时间、FPGA 硬件运行效果的影响。

2．实验内容

（1）single work item private 与 NDRange private。

参考 7.2.1 节的 kernel 代码，根据实验 10 的设计流程完成该实验内容。

（2）single work item local 与 single work item private。

参考 7.2.2 节的 kernel 代码，根据实验 10 的设计流程完成该实验内容。

（3）NDRange local 与 NDRange private。

参考 7.2.3 节的 kernel 代码，根据实验 10 的设计流程完成该实验内容。

（4）single work item local 与 NDRange local。

参考 7.2.4 节的 kernel 代码，根据实验 10 的设计流程完成该实验内容。

（5）single work item local float 与 single work item local char。

参考 7.2.5 节的 kernel 代码，根据实验 10 的设计流程完成该实验内容。

（6）NDRange private float 与 NDRange private char。

参考 7.2.6 节的 kernel 代码，根据实验 10 的设计流程完成该实验内容。

（7）single work item local float 与 ARM float。

参考 7.3.1 节的 kernel 代码，根据实验 10 的设计流程完成该实验内容。

（8）single work item local char 与 ARM char。

参考 7.3.2 节的 kernel 代码，根据实验 10 的设计流程完成该实验内容。

实验 12　具有一个隐形层的神经网络算法模型的 OpenCL 实现

1．实验目的

（1）分析比较 kernel 函数个数对算法实现的影响。

（2）kernel 函数之间 pipe 和 channel 通信机制的使用方法。

2．实验内容

（1）具有一个隐形层的神经网络的 TensorFlow 实现及训练。

参考 8.2 节的 TensorFlow 代码，根据实验 3 的实现步骤完成该实验内容。

（2）ARM 实现。

参考 8.3.1 节的 kernel 代码，根据实验 10 的设计流程完成该实验内容。

（3）single work item-local-float，一个 kernel 函数。

参考 8.3.2 节的 kernel 代码，根据实验 10 的设计流程完成该实验内容。

（4）NDRange-local-float，一个 kernel 函数。

参考 8.3.3 节的 kernel 代码，根据实验 10 的设计流程完成该实验内容。

（5）single work item-local-float，两个 kernel 函数。

参考 8.3.4 节的 kernel 代码，根据实验 10 的设计流程完成该实验内容。

（6）NDRange-local-float，两个 kernel 函数。

参考 8.3.5 节的 kernel 代码，根据实验 10 的设计流程完成该实验内容。

（7）single work item-local-float，两个 kernel，channel 通信。

参考 8.3.6 节的 kernel 代码，根据实验 10 的设计流程完成该实验内容。

（8）single work item-local-float，两个 kernel，pipe 通信。

参考 8.3.7 节的 kernel 代码，根据实验 10 的设计流程完成该实验内容。

实验 13　简易卷积神经网络算法模型的 OpenCL 实现

1. 实验目的

（1）掌握卷积神经网络的基本原理。

（2）分析比较 kernel 函数编程方式对算法实现的影响。

2. 实验内容

（1）简易卷积神经网络的 TensorFlow 实现及训练。

参考 9.2 节的 TensorFlow 代码，根据实验 3 的实现步骤完成该实验内容。

（2）NDRange 编程方式。

参考 9.3.1 节的 kernel 代码，根据实验 10 的设计流程完成该实验内容。

（3）single work item 编程方式。

参考 9.3.2 节的 kernel 代码，根据实验 10 的设计流程完成该实验内容。

参 考 文 献

[1] 陈仲铭,彭凌西.深度学习原理与实践[M].北京:人民邮电出版社,2018.
[2] 郑泽宇,梁博文,顾思宇.TensorFLow—实战 Google 深度学习框架[M].2 版.北京:电子工业出版社,2018.
[3] 王晓华.TensorFlow 深度学习应用实践[M].北京:清华大学出版社,2018.
[4] Hasitha Muthumala Waidyasooriya,Masanori Hariyama,Kunio Uchiyama. Design of FPGA-Based Computing Systems with OpenCL[M]. Cham Switzerland:Springer,2018.
[5] 黄乐天.FPGA 异构计算:基于 OpenCL 的开发方法[M].西安:西安电子科技大学出版社,2015.
[6] Intel. TensorFlow 培训教程[EB/OL].[2021/10]. https://software.intel.com/content/www/us/en/develop/training/course-applied-deep-learning-tensorflow.html.
[7] 友晶科技. DE10-Nano OpenCL User Manual[EB/OL].[2021/10]. http://www.terasic.com.cn/cgi-bin/page/archive_download.pl?Language=China&No=1048&FID=c164fa37e9283254f6e105f1b52888e2.